fusion

Sam Holyman
Phil Routledge
David Sang

Series Editor: Lawrie [...]

1

Nelson Thornes

Acknowledgements:

The Pupil Book authors wish to express their special thanks to the following people:

Neil Roscoe	Bev Routledge
John Payne	James Routledge
Ruth Miller	Joel Routledge
Jane Taylor	Jack Routledge
Judy Ryan	Jordan Routledge
Paul Lister	Sarah Ryan
Nick Pollock	Amanda Wilson
Darren Forbes	Annie Hamblin
Geoff Carr	Doug Thompson

Published in 2008 by:
Nelson Thornes Ltd
Delta Place
27 Bath Road
CHELTENHAM
GL53 7TH
United Kingdom

08 09 10 11 12 / 10 9 8 7 6 5 4 3 2 1

A catalogue record for this book is available from the British Library

ISBN 978 0 7487 9833 9

Illustrations by GreenGate Publishing, Barking Dog Art, Harry Venning and Roger Penwill

Cover photograph: Photo Library

Page make-up by GreenGate Publishing Services, Tonbridge, Kent

Printed and bound in the Netherlands by Wilco

Contents

Introduction

What is this book about?

This book is all about you. It aims to give **you** the ideas and knowledge to challenge the world around you. There is plenty of practical work to help you to understand the new ideas and to make it even more fun and interesting!

One of the great things about science is that it's all around you. We take so many things for granted around us – how our remote control works, where our food comes from, how our medicines help us to get better when we are ill, why salt and vinegar on chips tastes nice!

I don't care if you're conducting a scientific experiment, give me my chips back!

Next time you…

… run to catch the bus, make toast, see lightning, you can link your science knowledge to what you are seeing or doing. It's a great way to remind yourself of what you know.

Link up to…

other subjects! Science is one of those things that is found in lots of different areas, e.g. art and design, PE, geography, to name but a few

When you think about it, there are so many interesting questions that we don't have the answer to. This book will help you to look in the right place or investigate in the right way to find the answers. It will help you to see the connections between science and other subjects that you may be learning about at school.

Stretch Yourself

Some bits of science are easier to understand than others. If you find an area of science easy you might want to stretch yourself by trying some harder tasks

✚ Help Yourself

Some bits of science are difficult to understand and you may need a bit of help. This feature gives you tips and shows you connections that may make it all easier to understand.

activity

Activity

Science is a very practical subject and we have tried to make this book as active as possible. The activities in this book vary from full practical investigations to much smaller activities. They help to develop your knowledge and skills in as fun and active a way as possible.

But science doesn't have all the answers. That is what makes it so exciting and sometimes even a bit frustrating. It is a 'people' subject where scientists often work together, share ideas, discuss their opinions and present their findings.

There is a scientist in each and every one of us. If someone we know is ill, we may want to investigate what has caused their illness. When we hear that a face cream claims to get rid of spots, it is important that we understand how the company can support this claim. It helps to think like a scientist in your everyday life.

There are lots of features like 'Next time you'… and "Did you know?' in this book to help you.

Hmmm! Let me have a look at that!

SPOT CREAM

There is a scientist in all of us.

Did You Know?

There are lots of amazing facts about science. They are a great way to dazzle friends and family with your fantastic knowledge! Science certainly has the wow factor and sometimes it can be just plain gruesome.

Great Debates

Do you think you're right about everything? Do you enjoy getting into a heated discussion about day to day things, perhaps in the news or about your favourite music group? Being able to discuss and present an argument in science is an important skill. It's also a great way to find out more about interesting scientific issues.

science@work

Science is all around us in our everyday lives and in many people's jobs. Some jobs use science more obviously than others.

Questions in the yellow boxes. These are quick questions to check you understand the science you are learning.

Summary Questions

These check that you understand everything on a double page spread. Some questions are easy to answer. Others are more challenging and you may need to ask for help. You can use questions to help see what you understand well and to see where you can improve.

?

KEY WORDS

Important scientific words are shown in bold and appear in a list on the page and in the glossary at the back of the book.

Cells, Tissues and Organs

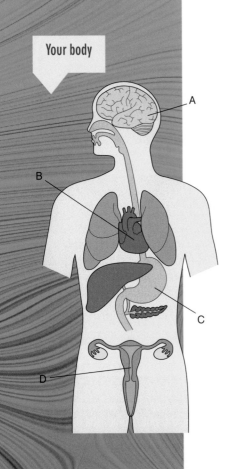

Your body

Your body

Your body is like a complicated machine. It is made of many organs, which work together. Organs are made from different tissues, each of which is made up of certain types of cell.

a Look at the diagram of the body. Four parts of the body are labelled A to D. Copy and complete the table below, choosing the name from the list below:

heart   lungs   stomach   intestines
kidney   liver   brain   uterus

Part	Name	Function
A		
B		
C		
D		

activity

Healthy heart?

- Measure your pulse rate when you are sitting down.
- Exercise for two minutes.
- Measure your pulse rate again.
- Get everyone in your class to do this and collect all of the results together.
- What should you do to ensure this is as fair a test as possible?
- What do the results tell you?
- Explain why this happens.

Fancy eating this?

Imagine if this was waiting for you at tea-time?

Yuk!

b What has happened to this food?

c What should you do to prevent this happening to food?

Food sometimes goes bad when microbes grow on it.

d Write down some more harmful things that microbes can do.

e Write down some ways in which microbes are useful.

activity

Having a closer look

Scientists often need to look at very small things, or they need to have a closer look at some big things. A magnifying glass can be very helpful.

- Use a magnifying glass to look at the skin on your fingertips. Draw what you see.
- Use your magnifying glass to look at the fabric of your school shirt or jumper. Draw what you see.
- What else can you find around the room to observe and draw?

Using a Microscope

» How do we focus a microscope?

» How much does a microscope magnify?

» How can we record what we see through the microscope?

What can we see when we look through a microscope?

Many parts of living things are too small to see. We need to **magnify** them so we can see them clearly. A magnifying glass can make things look about ten times bigger – not enough to see cells. For that we use a **microscope**, which has two lenses. The **specimen** we are looking at is magnified by each lens in turn. A microscope can make an **image** that looks hundreds of times bigger than the specimen.

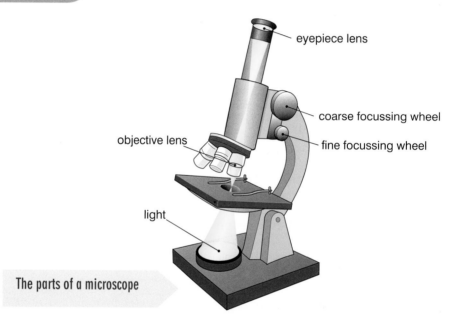

The parts of a microscope

Some microscopes have a lamp built in to them. Others have a mirror to reflect light from a separate lamp.

activity

Looking through a microscope

- Try looking at some of these objects through your microscope: a small piece torn from a newspaper, a human hair, a grain of salt or sugar.
- Put the object on a slide in the middle of the microscope stage.
- Use the lowest magnification **objective** lens.
- Turn the coarse focussing wheel so that the objective lens goes down as far as it can.
- Looking down the microscope, slowly turn the focussing wheel so that the lens moves up.
- Once you can see the specimen use the fine focussing wheel to make the image clearer.
- Try to draw what you can see.

Magnification

We can work out the magnification of a microscope so we know how much bigger the image is.

The eyepiece lens has a magnification number, such as ×5. This tells you how much that lens magnifies. Each objective lens also has a magnification number. Multiply the number on the **eyepiece** lens by the number on the objective lens. This tells you how much the specimen has been magnified.

> **a** You are using a ×5 eyepiece lens and a ×10 objective lens. What is the magnification?

> **b** What gives the biggest magnification, a ×6 eyepiece lens and a ×20 objective lens or a ×10 eyepiece lens and a ×10 objective lens?

Looking at cells

Hans Janssen and his son were Dutch spectacle makers. They found that by putting two or more lenses in a tube they could magnify very tiny objects. In 1590 they made the first microscope. Many scientists tried making microscopes. Anton van Leuwenhoek made microscopes with tiny lenses. He used them to observe blood cells and tiny water creatures.

Robert Hooke was an English scientist who made a microscope in 1665 and used it to look at a slice of cork. He noticed that there were lots of spaces in the cork, each one with a wall around it. He called these **cells** because they looked like the cells where monks lived. He drew what he saw and published his drawings in a book called *Micrographia*.

> **c** Who first used the word 'cell'?

Robert Hooke's drawing of cork cells

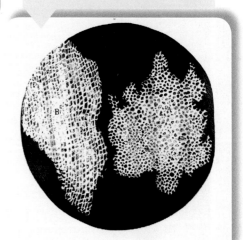

Summary Questions

1 Copy and complete:

Many parts of living things are too _____ to see, so we need to _____ them. A magnifying _____ can make things look about ten times bigger. To see even smaller things we need to use a _____. This can make things look _____ times bigger.

2 Write a set of instructions so that next year's new pupils will be able to use a microscope.

3 Why is it important to know what magnification we are using?

4 Use the Internet and books to find out more about Robert Hooke or Anton van Leuwenhoek. Make a poster or computer presentation to tell your class about what they discovered.

KEY WORDS

magnify
microscope
specimen
image
objective
eyepiece
cell

Looking at Animal Cells

» What parts of an animal cell can we see through a microscope?

» What are the jobs of the parts of an animal cell?

Making slides

We usually need to make a **slide** of our specimen so that we can look at it through a microscope. A slide is a piece of glass on which we put our sample. We usually cover the slide with a very thin piece of glass called a **cover slip**. This keeps the cells in place and stops them from drying out.

activity

Making a slide of animal cells

You are going to look at some of your own cells.

- Wipe the inside of your cheek with a cotton bud.
- Wipe the cotton bud onto a microscope slide. Dispose of the cotton bud as instructed.
- Put a drop of water on the slide. Put a cover slip on top.

 Safety: Follow your teacher's instructions about what to do with used cotton buds and slides. Wear eye protection when using disinfectant.

- Look at the slide under a microscope.

You might not be able to see very much when you look at your slide. We can help by adding a **stain**. This is a special dye that makes it easier to see things under a microscope.

- Carefully remove the cover slip.
- Add one drop of **methylene blue** stain.
- Look at the slide again.
- Try to draw what you see.

What are the parts of an animal cell?

Animal cells are difficult to see but a stain makes it easier. Cells are very complicated, but you will probably only be able to see three main parts:

- **Cell membrane** – this is a very thin layer that surrounds the cell, holding it together. It is a barrier which controls what substances get in and out of the cell.
- **Nucleus** – this controls the cell. It contains genes which carry information about how to make new cells.
- **Cytoplasm** – this is a jelly-like liquid which fills the cell. It has many different substances dissolved in it. Most of the cell's activities take place in the cytoplasm. The cytoplasm contains many different small structures. These are too small to see without an electron microscope, which magnifies much more.

Did You Know?

Your body contains about 1 000 000 000 000 cells. That's **one trillion** cells. If you counted two cells a second, it would take you 264 years to count all your cells!

The membrane of a cell is about 0.000 01 mm thick!

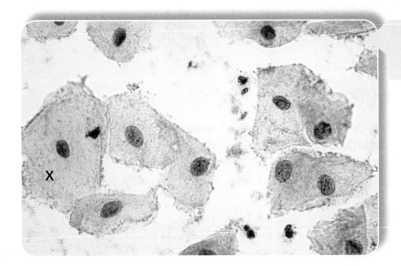

X

Human cheek cells. They have been stained to make it easier to see them

Human cheek cell

cell membrane

nucleus

cytoplasm

a The cells in the photograph have been magnified by about 2000 times. Measure the length of cell 'X' (in mm) and work out how big it is in real life. Show how you worked out the answer.

activity

Make a model cell

Make a model of an animal cell. Remember that the cell membrane is flexible. It is filled with cytoplasm, which is a watery jelly.

What could you use to make your model?

b Why do we use stains when we make a microscope slide?

Summary Questions

1 Copy and complete the table below:

Name of part	Function of part
Nucleus	
Cell membrane	
Cytoplasm	

2 What are the jobs of the cell membrane?

3 Which part of the cell contains genes?

4 What happens in the cytoplasm?

KEY WORDS

slide
cover slip
stain
methylene blue
cell membrane
nucleus
cytoplasm

Looking at Plant Cells

What parts of a plant cell can we see through a microscope?
How can we make a slide of plant cells?
What are the differences between animal and plant cells?

Know your onions

When you looked at animal cells you spread a very thin layer of cells on the slide. Onion cells are easy to look at through the microscope. That's because they have layers of **epidermis cells** which are very thin. You might have noticed them if you have ever peeled onions.

activity

Looking at onion cells

- Carefully use a scalpel to cut a layer of onion.
- On the inside of the layer you will find another very thin layer. This is the epidermis layer. Carefully peel it off using forceps/tweezers.
- On a white tile cut a piece of the epidermis about 1 cm × 1 cm.
- Carefully spread it on a microscope slide, keeping it as flat as possible.
- Add a drop of iodine solution. (CARE!) Wear eye protection.
- Put a cover slip on top.
- Observe your slide through a microscope.
- Draw what you can see.

> ⚠ **Safety:** Be careful!

Making a slide of onion cells

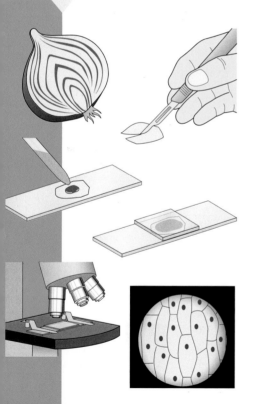

What is in a plant cell?

Plant cells have a cell membrane, a nucleus and cytoplasm. There are also some structures which we do not find in animal cells:

- **Cell wall** – this is made of a strong substance called **cellulose**. It is found outside the cell membrane. There are large spaces in the cell wall and substances can pass through it quite easily. It helps the cell to keep its shape.
- **Chloroplasts** – they contain a green substance called **chlorophyll**, which gives the plant its green colour. Chlorophyll collects light energy which the plant uses to make food.
- **Vacuole** – this is a space inside the cell. It contains a liquid called **cell sap**, which stores substances that the plant might need.

Onion epidermis cells

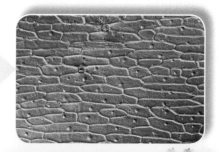

a Why do we have to make sure that the specimen on a slide is very thin?

b What did we add to the slide to make it easier to see the cells?

c Why do we put a cover slip on top of the specimen on a slide?

Chloroplasts in pondweed cells

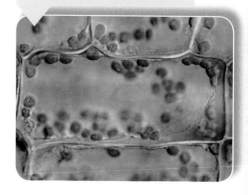

activity

Looking at green plant cells

Chlorophyll takes in light energy, which the plant uses for making food. Onion epidermis cells are found in the part of the plant that is underground. They do not have chloroplasts because light does not get to the epidermis cells.

We can see chloroplasts in small plants such as moss and pondweed.

● Put a small piece of the plant on a slide.
● Add a drop of water then put a cover slip on top.
● Squash the cover slip down gently.
● Look at the slide through your microscope.

Diagram of a typical plant cell

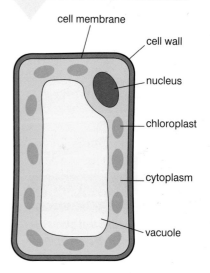

cell membrane
cell wall
nucleus
chloroplast
cytoplasm
vacuole

Summary Questions

1 Name three parts that are found in both animal and plant cells. Write a sentence to describe the job of each part.

2 Name three parts that are found in plants cells but not animal cells. Write a sentence to describe the job of each one.

3 Write two ways in which a cell membrane is different from a cell wall.

4 Do you think there are chloroplasts in root cells? Explain your answer.

5 Which part of a plant cell is made of cellulose?

6 Make a model of a plant cell. How will it be different from an animal cell?

KEY WORDS

epidermis cells
cell wall
cellulose
chloroplast
chlorophyll
vacuole
cell sap

Special Cells

Specialised cells

There are many different types of cells which have special jobs, or **functions**. **Specialised** cells have some of the same features, such as a membrane, cytoplasm and a nucleus, but they are not all identical. Different types of cell often have very different shapes and structures. Leaf cells have chloroplasts so they can absorb light energy to make food. Root cells are good at absorbing water from the soil.

The differences between types of cell make them good at doing a certain job. We say the cells are **adapted** for a particular function.

Sperm cells carry genetic information to an egg. They have a tail which they use for swimming. They have special structures which release the energy they need for swimming. The head of the sperm contains special chemicals that help it to penetrate an egg.

> **a** Why do sperm cells have a tail?

- ▶▶ What are the different functions of cells?
- ▶▶ How are cells adapted to carry out different functions?

Sperm cell

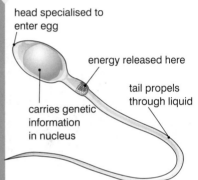

head specialised to enter egg

energy released here

tail propels through liquid

carries genetic information in nucleus

➕ Help Yourself

You should remember that plants use light energy to make their food. That is why there are lots of palisade cells packed tightly together in leaves. It means they can absorb as much light energy as possible.

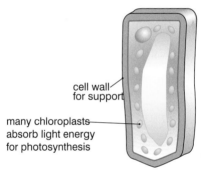

Palisade cells are part of the leaf that makes most food. They are packed with chloroplasts to absorb light. They are tall and narrow so that lots of palisade cells can fit close together.

cell wall for support

many chloroplasts absorb light energy for photosynthesis

Leaf palisade cell

> **b** Why are palisade cells long and narrow?

vacuole:
dissolved substances help draw water into root

Root hair cell

cell wall

large surface area:
helps efficient absorption

nucleus

Root hair cells absorb water and minerals from the soil. The 'hair' reaches far into the soil and gives the cell a large surface area, as so much of its surface is exposed. This means the cell can absorb lots of water for the plant.

> **c** What is the function of a root hair cell? Describe how it is adapted to carry out its function.

The main function of **red blood cells** is to carry oxygen from the lungs to other parts of the body. They are shaped to give them a large surface area so they can absorb oxygen more easily. The cytoplasm contains a chemical called 'haemoglobin', which carries oxygen.

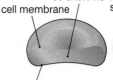

no nucleus, so short life

cell membrane

flexible shape to pass through small spaces

bi-concave shape gives large area to pick up oxygen

cytoplasm contains haemoglobin, which carries oxygen

Red blood cells

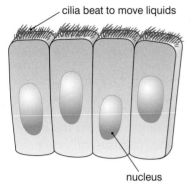

cilia beat to move liquids

nucleus

Lung epithelial cells

Our lungs are lined with **epithelial cells.** They are covered with tiny hair-like cilia. Cilia wave from side to side, brushing dirt and microbes from our lungs.

Nerve cells (or neurones) are very long so that they can carry messages to different parts of the body. They have many branches at the end so that they can connect with many other nerve cells.

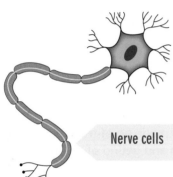

Nerve cells

Did You Know?

Red blood cells only live for about 120 days. After that they are broken down and useful bits are used to make new red blood cells. Two chemicals are made when red blood cells are broken down. They are called 'bilirubin' and 'biliverdin' and they give faeces (poo) its brown colour!

Summary Questions

1. Which type of cell has cilia?

2. Which cells contain haemoglobin?

3. Draw a table like the one below. Fill it in so that you can summarise the functions and adaptations of different types of cell.

Type of cell	Drawing of cell	Function of cell	Special features
Sperm cell			
Leaf palisade cell			
Root hair cell			
Red blood cell			
Lung epithelial cell			
Nerve cell			

4. Do some research to find out about one more type of specialised cell and add it to the table.

KEY WORDS

function
specialised
adapted
sperm cell
palisade cell
root hair cell
red blood cell
epithelial cell
nerve cell

Cells, Tissues, Organs and Systems

Tissues

Cells are adapted to carry out a particular function, but a cell does not work on its own. **Tissues** are groups of cells of the same type. They carry out a specialised function.

- Epithelial tissue lines the inside of the lungs. It keeps the lungs free of dirt and microbes.
- Muscle tissue is made of special cells that can contract (get shorter).
- Blood is a liquid tissue that carries substances around the body.
- Palisade tissue is found in leaves. It is the main site of photosynthesis in a plant.

Organs and systems

Your lungs contain special tissue which keeps them clean. They also contain other tissues. There are tissues which absorb oxygen, tissues which carry blood in and out of the lungs, and tissues which support the lungs. These different types of tissue work together to make the **organs** we call the lungs. The lungs work along with other organs to make the **respiratory system**. This is made up of the lungs and other parts of the body that take in oxygen and get rid of carbon dioxide.

The respiratory system

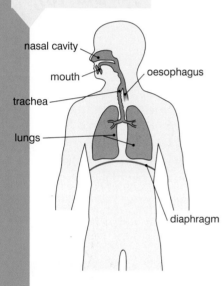

nasal cavity, mouth, oesophagus, trachea, lungs, diaphragm

The female reproductive system

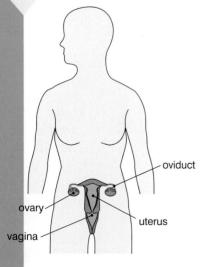

oviduct, ovary, uterus, vagina

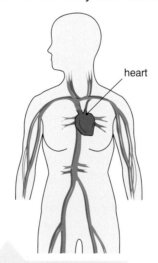

heart

The circulatory system

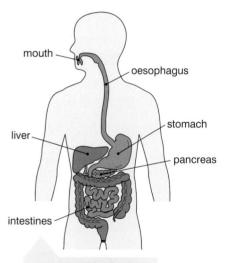

mouth, oesophagus, stomach, pancreas, liver, intestines

The digestive system

The **circulatory system** carries blood around the body. Blood is pumped by the heart. It is carried in blood vessels.

The **digestive system** is made up of the mouth, gullet, stomach and intestines. Food gets broken down then absorbed into the blood.

The **reproductive system** is made up of the organs used to make babies. Unlike other organ systems, a woman's reproductive system is totally different from a man's.

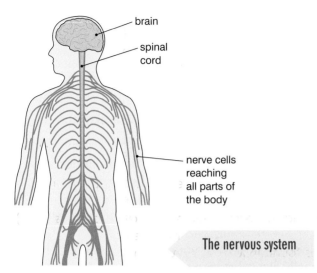

brain

spinal cord

nerve cells reaching all parts of the body

The nervous system

No, wrong sort of organ!

The **nervous system** is made up of the brain and spine, which connect to the rest of the body through nerves.

activity

Your organ systems

You need a very large sheet of paper.

● Lie down on the paper and get your partner to draw the outline of your body.

● Use this page, and any other sources of information, to draw the organs that make up your different body systems.

● Try to make sure that each organ is in the correct place.

● Label your drawing or make a key to identify the different parts.

● Name some organ systems that are not mentioned above.

Link up to...

ART OR DESIGN & TECHNOLOGY

Make a t-shirt showing the size and positions of the organs in your chest and abdomen. This will help you to remember where the organs are found and might make a great art project!

Summary Questions

❶ Complete each of these sentences:
a) The function of the respiratory system is …
b) The main parts of the circulatory system are …
c) The reproductive system of a woman …
d) The nervous system is important because …

❷ Some parts of the body can be replaced if they go wrong. Write a list of parts that can be transplanted from another person.

KEY WORDS

tissue
organ
respiratory system
circulatory system
digestive system
reproductive system
nervous system

The Skeleton

What are the functions of our skeleton?

What makes bones strong?

Did You Know?

We have 206 bones in our body. The smallest ones are inside our ears!

Finding out about bones

The **skeleton** is one of the systems of our body. The skeleton is made of bones. It has four main jobs:

- It allows us to move when muscles pull our bones. Bones are held together at joints, which allow bones to move.
- It supports our body.
- It protects our soft internal organs, such as the brain, heart and lungs.
- It makes new blood cells. Most of your bones are hollow. They are filled with a jelly-like substance called **marrow**. This is where most of our blood cells are made.

activity

Investigating bone

Bone is made of a mixture of calcium salts and protein. These substances combine together to make a very strong material. We can investigate what happens if we remove one of these substances.

Removing the calcium salts

- Get a bone from a chicken leg.
- Put it in a beaker of dilute acid and leave it overnight. This will dissolve the calcium salts. Wear eye protection.
- Use a pair of tongs to remove the bone from the acid. Rinse the bone in cold water.
- What do you notice about the bone now?

Removing the protein

- Put another piece of bone in a crucible and heat it with a Bunsen burner for about 10 minutes. Wear eye protection. This will break down the protein in the bones.
- Turn off the Bunsen and let the bone cool.
- What do you notice about the bone?

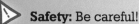 **Safety:** Be careful!

Strong bones

Bones are made of a framework of protein, hardened by the calcium salts. This is why calcium is an important part of our diet, especially in people who are still growing.

The human skeleton

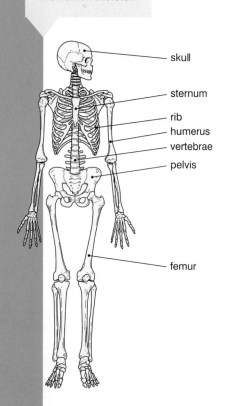

- skull
- sternum
- rib
- humerus
- vertebrae
- pelvis
- femur

Vitamin D is also important, because it helps our bodies to absorb calcium from food. People who lack vitamin D or calcium may suffer from rickets, where the bones bend under the weight of the body.

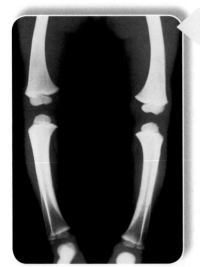

X-ray of a child with rickets

🎵The thigh bone's connected to the hip bone... 🎵

Stretch Yourself

Find out about how **chiropractors** and **osteopaths** use their knowledge of the skeleton to help relieve aches and pains.

a Which mineral and which vitamin are needed to make healthy bones?

activity

Hollow bones

Bones are hollow and are filled with marrow. Is a hollow bone weaker than a solid one? Your teacher might show you an animal bone from the butcher.

- Get a sheet of A4 paper and roll it into a cylinder about 1 cm in diameter.
- Hang weights on it until it bends. Protect the bench and feet from falling weights.
- Record the mass it supported.
- Use another sheet of A4 paper to make a solid rod by rolling it up as tightly as you can.
- Test it to find out what mass it supports.
- Which is stronger, a hollow tube or a solid rod of the same mass?

Summary Questions

1 Which part of our body is protected by our skull?

2 What important organs are protected by our ribs?

3 What is the job of bone marrow?

4 Why are bones hollow?

KEY WORDS

skeleton
marrow

Joints and Muscles

- ▸▸ How does our skeleton move?
- ▸▸ How do joints work?

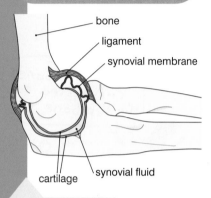

A synovial joint

The ball and socket can be clearly seen in this hip joint

The elbow is a hinge joint

Joints

Our bones are held together by strong, stretchy **ligaments**. The ends of the bones are covered with a layer of **cartilage**. This is slippery and flexible. It acts as a shock absorber and allows bones to move freely. It prevents the ends of bones from rubbing together.

The space between the bones is filled with **synovial fluid**. This lubricates the joint, like the oil on your bike chain.

a Explain how a synovial joint is able to move smoothly.

b 'Osteoarthritis' is a condition where the cartilage in joints becomes rough and may break away from the bones. What symptoms would an osteoarthritis sufferer have?

Types of joint

The place where two bones meet is called a 'joint'. We have different types of joint that move in different ways:

Ball and socket joint – this is the type of joint in our shoulder and hip. It gives our arms and legs a wide range of movement.

Pivot joint – this is the type of joint where our skull joins our spine. It gives us limited movement in all directions.

Hinge joint – this is the type of joint in the elbow and knee. It only allows movement in one direction, like a door hinge.

Fixed joint – our skull is made of several bones. When we were born, these bones were not joined together. As we get older they join together, forming fixed joints.

c Which types of joint can be replaced by artificial ones in operations?

d Why are the bones of our skull not joined together when we are born?

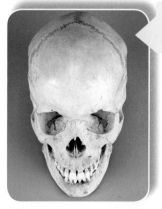

The fixed joints can be seen on top of the skull

Joints

Your teacher will give you an animal joint to examine.

- Describe features you can see that make sure the joint moves smoothly.
- Describe features of the joint that stop the bones from knocking together when the animal is moving.
- Wash your hands after handling bones.

Muscles

- Muscle keeps our heart beating.
- Muscle pushes food through our guts.
- Muscle pumps blood through our arteries.
- Muscle allows us to move our skeleton so we can walk and move around.
- Muscle helps us to breathe by moving our rib cage up and down (skeletal muscle).

Muscle cells can **contract** (get shorter and thicker). When they finish contracting they **relax**. If a relaxed muscle gets pulled by another muscle it will return to its original length and thickness. Muscles are joined to bones by strong **tendons**. Tendons are **inflexible** – they do not stretch.

Muscles move bones by pulling them. Muscles cannot push bones. Muscles work in pairs. One muscle bends the joint and the other straightens the joint. Look at the diagram of the arm.

e What will happen when the biceps contracts?

f What will happen when the triceps contracts?

science @ work

Physiotherapists use their knowledge of bones and muscles to help patients recover from injuries.

Muscles in the arm

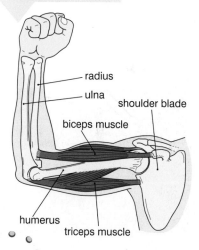

radius
ulna
shoulder blade
biceps muscle
humerus
triceps muscle

Summary Questions

1 Complete the table to summarise what you have learnt about the different types of joint:

Type of joint	Example	Range of movement
Ball and socket	Hip, shoulder	Allows a wide range of movement in all directions
Hinge		
Pivot		
Fixed		

2 Why can muscles pull bones but not push them?

3 Why do ligaments need to be able to stretch?

4 What could happen if the ligaments stretch too much?

5 Why is it important that tendons are not able to stretch?

KEY WORDS

ligament
cartilage
synovial fluid
ball and socket joint
pivot joint
hinge joint
fixed joint
contract
relax
tendon
inflexible

Microbes

▸▸ What are the different types of microbe?

▸▸ How can we see microbes?

▸▸ What do microbes look like?

Microbes, or **micro-organisms**, are all around us. We have microbes on our skin and even inside our bodies. Most microbes have no effect on us at all. Some microbes cause diseases, while others are very useful in many different ways.

There are four main types of microbe:

- Fungi
- Protozoa
- Bacteria
- Viruses

Fungi

There are many different types of **fungus**. Some, such as yeast, are single celled. We can only see them by using a microscope. Most fungi live as fine threads that grow through soil. Fungi decay dead animal and plant material, recycling nutrients from them. Most fungi reproduce by making spores. Mushrooms and toadstools are the spore-producing parts of a fungus.

> **a** Are all fungi microscopic? Explain your answer.

Bacteria

There are thousands of different species of **bacteria**. They are just big enough to see through an ordinary microscope. They are found in soil, in water and even inside living animals. Bacteria are single-celled organisms and most are either round or 'sausage'-shaped. They reproduce by splitting in two.

Some species of bacteria have a 'tail' (flagellum). They use this to help them move around.

Some bacteria feed on dead animal and plant material, making it decay. This recycles nutrients.

> **b** What do bacteria use their flagellum for?

➕ Help Yourself

You will sometimes see the word 'germ' used instead of 'microbe'. Germ is not a very scientific word and if you write it in an exam it will probably be marked wrong!

Fungi

Did You Know?

We have bacteria living in our intestines. They produce many of the gases that are released when we pass wind, including the smell!

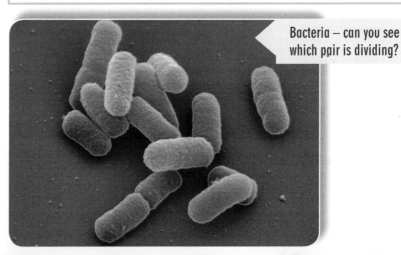

Bacteria – can you see which pair is dividing?

Protozoa

Protozoa are single-celled organisms. Some of them are just big enough to see with the naked eye. Many can be seen clearly with a magnifying glass. They mostly live in ponds or in very damp places. They feed on bacteria and other smaller microbes, by engulfing them and then digesting them.

c How do protozoa feed?

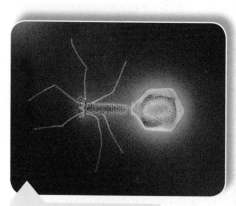

One species of protozoa (an amoeba) eating a smaller one (a paramecium)

Viruses

Viruses are the smallest microbes. We can only see them with an **electron microscope**.

They can only reproduce by taking over the cells of other organisms. Viruses can infect animals, plants and even bacteria. Once inside a cell they turn it into a 'virus factory'. The cell makes new viruses which then infect other cells. Most scientists do not count them as living things because they can only reproduce inside other cells.

d Why do some scientists say that viruses are not alive?

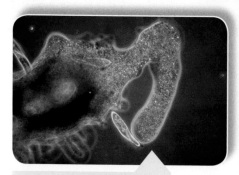

T4 virus infects bacteria

Looking at microbes

Your teacher will give you some samples of microbes to look at. Start by looking with a hand lens. Describe what you can see.

- Examine the specimens with a microscope. Your teacher will show you how to prepare slides to look at.
- Draw what you see through the microscope.
- Calculate the magnification you used and write it on your drawings.

Summary Questions

1 How are viruses different from other microbes?

2 List the four types of microbe in order of size, starting with the smallest.

3 Which microbes recycle useful nutrients by decaying dead animals and plants?

4 Draw a table to summarise the differences between the four types of microbe. Include information about how they feed, their structure and how they reproduce.

KEY WORDS

microbe
micro-organism
fungus
bacteria
protozoa
virus
electron microscope

Growing Microbes

We can grow bacteria and fungi quite easily. It is much more difficult to grow viruses, as they can only reproduce in living cells. Scientists have to make cultures of human or animal cells and grow viruses in the cells.

» How can we grow bacteria?

» How can we measure factors that affect the growth of yeast?

Growing bacteria

Most bacteria will grow and reproduce if they have the nutrients they need and are kept in a warm place. We usually grow bacteria in a special nutrient broth or jelly. They are often grown on **nutrient agar jelly** in a special plastic **Petri dish**.

Scientists must take great care when growing bacteria. If a few harmful bacteria got onto the nutrient jelly, there could be hundreds of millions of them in a couple of days. Scientists have to use **aseptic** techniques when growing bacteria. This means that they have to make sure that everything is sterilised before they start. They must make sure that no stray bacteria can get onto the nutrient jelly.

After a few days the bacteria will have reproduced. They form **colonies** of millions of bacteria. The colonies are big enough to see.

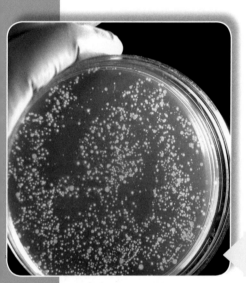

Colonies of bacteria can be seen on the nutrient agar

Growing bacteria

activity

- You will need two Petri dishes containing nutrient agar.
- Take the lid off the dish for as short a time as possible.
- Take care not to contaminate the dish and do not touch the jelly with your fingers. You must never spit or put any bodily fluids on to nutrient agar jelly.
- Your teacher will show you how to transfer bacteria onto the nutrient agar jelly. This is called **inoculation**.
- Use two short pieces of Sellotape to hold the lid on the Petri dish. Do **not** seal the dish with Sellotape.
- One of the dishes will be **incubated** in a cold place and the other will be kept in a warm place.
- After the dishes have been incubated for a couple of days you will be able to examine them.
- How does the temperature affect the growth of bacteria?

 Safety: Do not remove the Sellotape or open the dishes.

ⓐ Why must the area where you are working be sterilised before inoculating your nutrient agar?

ⓑ Why must you never touch the nutrient agar jelly?

Growing yeast

You probably know that yeast is used in bread making. Yeast uses sugar as a source of energy and produces carbon dioxide. Carbon dioxide forms bubbles in the dough. When the dough is baked the bubbles stay in the bread, giving it a spongy texture.

activity

What factors affect the rate of respiration of yeast?

- Mix 20 g of flour with 1 g of sugar.
- Add 20 cm³ of warm yeast mixture.
- Pour the mixture into a 100 cm³ measuring cylinder.
- Record the volume of the mixture every 2 minutes for 30 minutes.
- What happened to the level of the mixture in the experiment?
- Why does this happen?
- What factors could affect the rate of respiration in yeast? Choose one factor to investigate. Explain how you will carry out your investigation.
- Use your scientific knowledge to predict what you think will happen.

Now do your investigation.

measuring cylinder

glass rod

Flour

beaker

flour

yeast

SUGAR

sugar

stop watch

Growing microbes

Summary Questions

❶ What kind of dish did you use for growing bacteria?

❷ What does nutrient agar jelly provide for the bacteria?

❸ Make an information sheet to tell other pupils how to grow bacteria safely. For each instruction you write, make sure you explain why it is important.

KEY WORDS

nutrient agar jelly
Petri dish
aseptic
colony
inoculation
incubate

Useful Microbes

>> How are microbes used to make foods and other products?

Many microbes are useful. Some fungi and bacteria have been used to make food for thousands of years.

Yeast

You already know that yeast is used to make bread. As the yeast respires in the dough, it produces carbon dioxide. This makes the dough rise and gives the bread a light texture. When yeast does not have any oxygen it produces some ethanol (or alcohol) as it respires. When the bread is baked the alcohol evaporates.

Yeast is also used to make beer and wine. This is called **fermentation**. Yeast is mixed with something that contains sugar. Grape juice is used to make wine and apples are used to make cider. Barley is a type of cereal. It contains a lot of starch. When the barley starts to grow, the starch is turned to sugar. This makes malted barley, which is used to make beer. When we make wine, the carbon dioxide escapes into the air; but in beer and champagne it is trapped, giving the drinks their 'fizz'.

Even when the beer is made, the yeast carries on being useful. Some of the yeast is used to make the next lot of beer, but a lot of it is left over and can be used to make yeast extract, like Marmite. This is a yeast extract, which is a good source of vitamin B.

Products made from yeast

> **a** Draw a flow chart to explain how beer is made.

activity

Yoghurt

If you do this in a food technology room it will be safe to eat the yoghurt you make.

- Boil 500 cm³ of milk then leave it to cool in a clean container.
- Add two teaspoons of natural yoghurt. This contains live *Lactobacillus* bacteria.
- If possible, put a pH sensor in the mixture and connect it to a data logger. If it is not possible then take out a small sample of the mixture. Use universal indicator to test its pH.
- Cover the mixture and leave it in a warm place overnight.
- Either download and look at the data for pH, or take out another small sample and test it again.
- Why was the milk boiled and why was it left to cool?
- Why was the mixture kept in a warm place?
- What happened to the pH and the thickness of the mixture overnight?

Milk contains a sugar called 'lactose'. Bacteria use this as a source of energy. They make a substance called **lactic acid**, which affects the protein in the milk. The change in the protein makes the milk thicken, turning it into yoghurt.

Cheese

Rennet is extracted from the stomachs of calves. When it is added to milk it changes the milk to a solid part called **curd** and a liquid part called **whey**. To make cheese, the curd is collected and compressed and a mixture of bacteria and fungi is added.

As well as using microbes to make foods humans have eaten many fungi, like mushrooms, for thousands of years. There are also new foods made from fungi. Waste potato starch from making crisps is used as a source of energy for growing a fungus called *Fusarium*. This is then made into a high protein food called 'Quorn'. This is flavoured and used to make 'meat-like' products that are low in fat and acceptable to vegetarians.

... eat a carton of yoghurt, just remember you are eating milk which has been changed into a solid by lactic acid, excreted by bacteria!

activity

Mouldy cheese

Some cheeses, called blue cheeses, have moulds added to them.

- Use a microscope to look at the mould from a blue cheese.
- Draw what you see.

I'm not sure if I want to eat this after all!

Biotechnology

Biotechnology means using living things to make new products. In recent years there has been a revolution in biotechnology with many new products:

- Microbes make **enzymes** that are used in washing powders to remove stains.
- Microbes make enzymes that turn waste starch into sugar syrup used in fizzy drinks.
- Microbes can be used to convert sewage into methane gas that can be used as a fuel.

- Microbes convert sugar cane into alcohol, which is used as a renewable fuel for cars.
- Microbes are used to make antibiotics, such as penicillin, that are used to treat diseases.
- Microbes make medicines such as insulin, which is used to treat diabetes.

Cheeses contain a mixture of bacteria and fungi

Summary Questions

1. All cheeses are made of the same ingredient, milk, so how are there so many different cheeses?

2. Use the Internet or books to find out what an enzyme is.

3. Oil is likely to run out in the next 50 years. How could microbes help solve this problem?

4. Make a poster showing how biotechnology affects our lives. Remember to include the 'old' biotechnology such as bread making and brewing.

KEY WORDS

fermentation
lactic acid
curd
whey
biotechnology
enzyme

Harmful Microbes

- ▶▶ What diseases are spread by microbes?
- ▶▶ How are diseases spread?
- ▶▶ How does our body defend itself against microbes?

The chickenpox virus infects skin cells

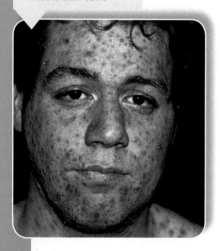

I should win this easily - my doctor says I've got athlete's foot!

We are surrounded by microbes. Most of them do us no harm and some are very useful. Some make food 'go off' and make milk go sour. Some microbes cause **diseases**. Microbes which cause diseases are called **pathogens**. Diseases which can be passed from person to person are called **infectious** diseases. The table tells you about some of the diseases caused by different microbes:

Type of microbe	Diseases which they cause
Viruses	Common cold, flu, measles, chickenpox, AIDS
Bacteria	Food poisoning, cholera, tuberculosis (TB), impetigo
Protozoa	Malaria, sleeping sickness
Fungi	Athlete's foot, thrush, ringworm

ⓐ Which type of microbe causes malaria and which type of microbe causes measles?

Spreading diseases

Diseases are spread in several different ways:

- **Droplets** – diseases like the common cold and flu make us cough and sneeze. Tiny droplets carry the pathogen in the air and are breathed in by someone else.
- **Food and water** – some pathogens are taken in when we eat and drink. Contaminated food can pass on salmonella. Contaminated water can pass on cholera and typhoid.
- **Animals** – several diseases can be spread by animals. They include malaria, which is passed on by mosquitoes; Black Death, which is passed on by fleas living on rats; and rabies, which is passed on in animal bites.
- **Skin** – some diseases are passed on when you touch an infected person, or even touch something they have touched. Impetigo is often passed on by using the same towel as an infected person. Walking on a wet changing room floor, where an infected person has walked, can pass on athlete's foot.
- **Blood** – diseases such as AIDS and hepatitis can be passed from person to person using the same needle to inject drugs.
- **Sex** – some diseases, such as syphilis and AIDS, can be passed on through sexual contact. Using a condom during sexual intercourse can greatly reduce the risk of catching an infection (i.e. a sexually transmitted infection, an STI).

Barriers to infection

Our bodies have barriers, such as skin, to keep microbes out.

Sometimes microbes get past these barriers shown above. **White blood cells** make the next line of defence. Some white blood cells 'engulf' and digest microbes. Other white blood cells make special chemicals called **antibodies**. These make microbes stick together so it is easier to engulf them and also stops them from reproducing. Babies get some antibodies from their mother while still in the womb and in the first breast milk.

b What are cilia and how do they prevent lung infections?

c How does your body stop cuts from getting infected?

Keeping microbes out

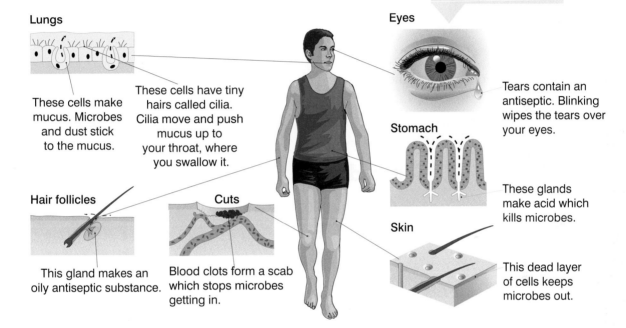

Lungs

These cells make mucus. Microbes and dust stick to the mucus.

These cells have tiny hairs called cilia. Cilia move and push mucus up to your throat, where you swallow it.

Hair follicles

This gland makes an oily antiseptic substance.

Cuts

Blood clots form a scab which stops microbes getting in.

Eyes

Tears contain an antiseptic. Blinking wipes the tears over your eyes.

Stomach

These glands make acid which kills microbes.

Skin

This dead layer of cells keeps microbes out.

1 Is it true that 'coughs and sneezes spread diseases'? Explain your answer.

2 What is meant by a 'pathogen'?

3 Complete the table to summarise the ways that infectious diseases can be passed on:

How microbes are passed on		Examples
Droplet	When we cough or sneeze tiny droplets in the air carry pathogens, which can then be breathed in by someone else.	Common cold, flu
Food and drink		
Animals		
Blood		
Skin		
Sex		

Summary Questions

KEY WORDS

disease
pathogen
infectious
white blood cell
antibody

know your stuff

B1

▼ Question 1 (level 3)

Jack is having an X-ray of his arm.

The photo below shows what the X-ray looks like.

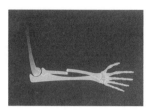

a Which parts of Jack's arm showed up on the X-ray? [1]

b Look at the drawing of the X-ray. What has happened to Jack's arm? [1]

c The diagram below shows Jack's arm.

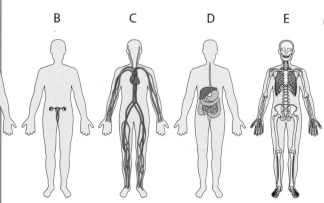

(i) Make a rough sketch of the diagram. Then draw a line from the letter J to a joint in the arm. [1]

(ii) Why do our arms and legs have joints? [1]

(iii) Name the parts that make the arm move. [1]

▼ Question 2 (level 4)

The diagram below shows some organ systems in the human body.

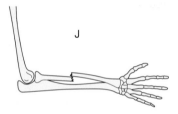

a Name the organ systems shown by each letter in each diagram. Choose from the list below:

circulatory system

skeleton

digestive system

respiratory system

reproductive system [5]

b Which one of these organ systems is totally different in a male and a female? [1]

▼ Question 3 (level 6)

The diagram shows a plant cell.

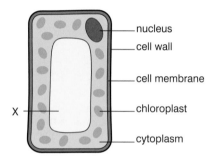

a Which two named parts do we find in plant cells but not animal cells? [2]

b Which named part contains information about how to make new cells? [1]

c Which named part absorbs light to make food in photosynthesis? [1]

d Name the part labelled X. [1]

e Where in a plant would you find a cell like the one in the diagram? [1]

How Science Works

▼ Question 1 (level 5)

Bacteria in the skin can cause spots. A scientist investigated the effect of spot cream on bacteria.

a He grew bacteria on the surface of jelly in a Petri dish.

At what temperature would the bacteria reproduce quickly?

Choose from the temperatures below:

90°C   10°C

35°C   −10°C   [1]

b The researcher placed two small circles of paper onto the surface of the jelly.
One had been soaked in spot cream. The other had been soaked in water.
The diagrams below show the jelly at the start of the experiment and two days later.

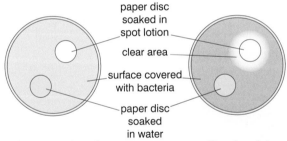

paper disc soaked in spot lotion

clear area

surface covered with bacteria

paper disc soaked in water

At the beginning of the experiment   **Two days later**

What happened to the bacteria in the clear area around the paper soaked in spot cream?   [1]

c What was the control used in this experiment?   [1]

d What safety precautions should the researcher take in this investigation?   [2]

▼ Question 2 (level 6)

One evening Sitel and Runi ate chicken salad which had been left in Runi's locker all day. The next day both girls were ill. Their doctor gave them antibiotics to take for eight days.

The graph shows the effect of antibiotics on the number of bacteria in the body.

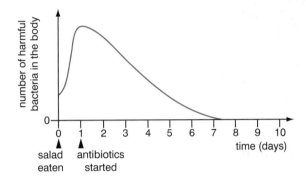

a Use the graph to explain why the girls did not become ill until the day after eating the salad.   [1]

b After taking the antibiotics for eight days, Runi was completely better. Explain why she got better.   [1]

c Sitel was told to take the antibiotics for eight days. She felt much better after five days so she stopped taking the antibiotics. The next day she felt very ill again.
Use the graph to help you explain why Sitel became ill again.   [2]

d Food will keep longer if you place it in a refrigerator at 4°C.

This does not kill bacteria.

What is the effect of low temperature on bacteria?   [1]

B2.1 Reproduction

We'll have omelette for tea. A dozen hens' ova please.

Fertilisation

- Animals and plants reproduce to make new **offspring**. Animals mainly use **sexual reproduction**. This needs two animals, a male and a female.

- All animals make sex cells or **gametes**. Male gametes are called sperm cells. Female gametes are called eggs. A more scientific name for an egg is an **ovum**. The plural of ovum is **ova**.

- Each gamete contains a nucleus. The nucleus of the sperm has to join together with the nucleus of the egg. This is called **fertilisation**. When an egg has been fertilised, it can start to grow into a new animal.

- Before fertilisation can happen, the sperm has to get to the eggs. There are several ways that this can happen:

  - External fertilisation
  - Internal fertilisation

Stickleback eggs are externally fertilised

External fertilisation

- Many water animals use external fertilisation (where fertilisation takes place outside the body). The female lays eggs and the male squirts sperm on top of them. Male sticklebacks make a nest. The female lays eggs in the nest and then the male fertilises them.

- Male frogs hold onto the female and squirt sperm on the eggs as the female lays them.

- Sperm does not reach all the eggs. A lot of them will not be fertilised. Animals with external fertilisation make lots of eggs. This makes sure that at least some will be fertilised.

Did You Know?

The Ocean Sunfish makes 300 000 000 eggs at a time!

Internal fertilisation

- Internal fertilisation (where fertilisation takes place inside the body) increases the chance of an egg being fertilised. So animals with internal fertilisation make fewer eggs.

- In some animals, such as mammals, the fertilised egg develops inside the mother. In others, such as birds and most reptiles, the female lays eggs. The egg develops externally. Some reptiles lay eggs and leave them to develop on their own. Other reptiles protect their eggs.

Birds eggs are internally fertilised

activity

Comparing fertilisation

Copy this table:

	External fertilisation	Internal fertilisation
External development		
Internal development		

Write these animals in the correct spaces: human, frog, goldfish, lizard, blackbird, horse, newt, turtle, salmon.

- Use the Internet or books to find some animals that do not fit in with others in their group. For example, some fish have internal fertilisation and development.

Caring for the young

- Many animals do not look after their fertilised eggs. They hatch and the young animals look after themselves. Many of them get eaten or starve to death. Animals that do not look after their young lay lots of eggs. Other animals look after their young once they are born or hatch from an egg. Mammals even make a special food to feed babies – milk.

Mammals make milk to feed their young

Birds bring food to feed their young

Sex Organs

▶▶ What do the different parts of a woman's sex organs do?

▶▶ What do the different parts of a man's sex organs do?

A new life begins when an egg is fertilised by a sperm. Females make eggs and males make sperm. Our sex organs are the bits of our body that make eggs and sperm, and help them to join together.

Name of part	What it does
Ovary	Makes eggs and releases one every month. Also makes female sex hormones.
Uterus	Where a baby grows when a woman is pregnant. It is sometimes called the 'womb'. The uterus has walls made of strong muscles.
Oviduct	Carries an egg from the ovary to the uterus. This is where sperm meet an egg. It is lined with small 'hairs' called cilia. These push the egg along.
Vagina	Where the penis fits into a woman and where sperm go when they leave the man's penis.
Cervix	The narrow opening from the vagina to the uterus.

Here are diagrams of the female and male sex organs:

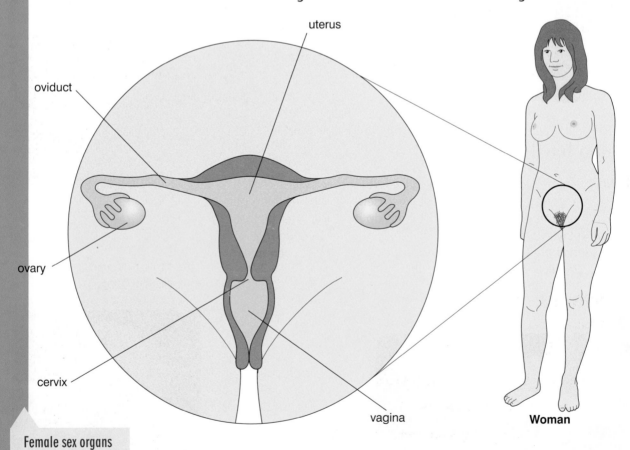

uterus

oviduct

ovary

cervix

vagina

Woman

Female sex organs

Male sex organs

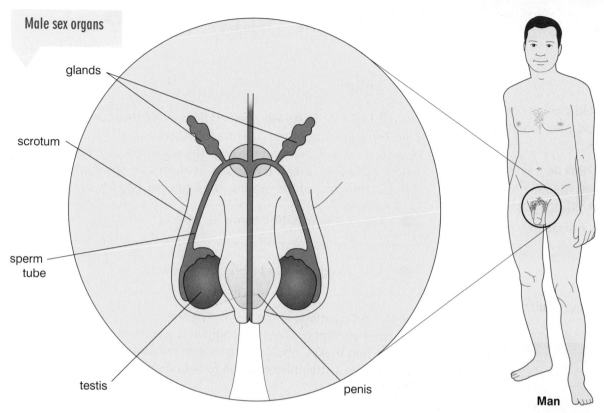

glands

scrotum

sperm tube

testis

penis

Man

Name of part	What it does
Testis (plural: testes)	Makes sperm. Also makes the male sex hormone called 'testosterone'.
Penis	Fits inside the vagina. Where the sperm come out during ejaculation.
Glands	Adds liquid to the sperm to make semen.
Scrotum	Bag that holds the testes outside the body. This keeps them at the right temperature to make sperm. This is slightly lower than body temperature.
Sperm tube	Carries sperm from the testes to the penis.

Did You Know?

When a girl is born, her ovaries already contain every egg she will ever make. A woman will release about 450 eggs in her whole life. A man releases about 250 million sperm each time he has sex! A man can make sperm continuously from puberty to death and this is why older men can become fathers.

Summary Questions

1 Sort out these words into two columns, one headed 'male' and one headed 'female': **penis, vagina, testes, ovaries, scrotum, uterus, sperm, egg, oviduct, sperm tube.**

2 What is the job of the ovaries?

3 What is the job of the testes?

4 How often does a woman produce an egg?

5 How do eggs move into the uterus?

6 Why are the testes in the scrotum outside the body?

KEY WORDS

ovary
uterus
oviduct
vagina
cervix
testis (plural: testes)
penis
glands
scrotum
sperm tube

Fertilisation in Humans

▸▸ How are sex cells adapted to do their jobs?

▸▸ How do a sperm and an egg get together?

▸▸ What happens in fertilisation?

Sex cells

The photograph shows a magnified human egg surrounded by sperm cells. The egg is about the size of a full stop on this page. The sperm cells are much smaller than the egg. The egg contains enough food to last a few days after it is released. Every month one egg is normally released from one of the ovaries. The egg goes into the oviduct.

a What is the scientific name for an egg?

b How are eggs moved along the oviduct?

Sperm cells are made in the testes. The sperm have a tail that is used to push them along. The head contains a special chemical called an enzyme that makes a hole in the membrane (outside) of the egg. Sperms have a streamlined shape to help them swim to the egg.

Human sperm fertilising an egg

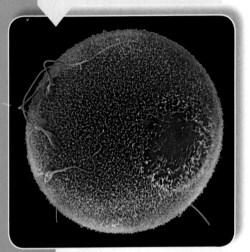

Egg and sperm

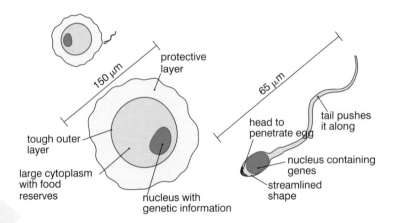

protective layer

150 μm

tough outer layer

large cytoplasm with food reserves

nucleus with genetic information

65 μm

head to penetrate egg

tail pushes it along

nucleus containing genes

streamlined shape

Sexual intercourse

The penis is usually soft. When a man is ready to have sexual intercourse extra blood is pumped into the penis. This makes his penis get bigger and harder. This is called an **erection**.

When a woman is ready to have sexual intercourse her vagina gets wider and makes extra liquid. This lubricates the vagina so the penis can move in easily. The man and woman move so that the penis slides in and out of the vagina. This gives them lots of pleasure. Eventually muscles contract, pumping sperm from the testes. This is called an orgasm. Liquid is added from the glands. This makes a mixture called **semen**, which travels into the vagina. The release of semen is called **ejaculation**.

Fertilisation

Once the sperm cells are inside the vagina they swim through the cervix and into the uterus. Meanwhile the woman may have recently released an egg. Some of the sperm will swim into the oviduct and surround the egg. One sperm might manage to get into the egg by making a special chemical that breaks down the membrane. The tail of the sperm breaks off. The sperm nucleus joins together, or fuses, with the egg nucleus. This is **fertilisation**.

Fertilising an egg is not the only reason why people have sexual intercourse. Men and women have sex as a way of showing their love and giving pleasure to each other – that is why it is called 'making love'. If a couple want to have sex without fertilising an egg they will need to use **contraception**. You will learn more about contraception later in the topic.

Courtship

Many animals perform **courtship** displays to attract a mate. Female peafowl are attracted to the male with the biggest tail. You have probably noticed humans courting members of the opposite sex!

Apparently its called a courtship ritual.

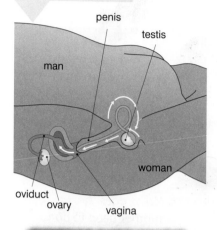

Sexual intercourse

+ Help Yourself

Find out about how features of the mother and father are passed on to their child. Look at spread B1.3 'Looking at Animal Cells'.

The peacock's tail helps it to attract a mate

Summary Questions

❶ Copy and complete:

Sperm are made in the _____. When a man ejaculates the sperm mixes with liquid to make _____. The sperm are released into the _____ and swim through the _____ into the _____. Eventually some sperm cells may meet an egg in the _____. The nucleus from one sperm might fuse with the egg nucleus. This is called _____.

❷ How are eggs and sperm adapted to carry out their jobs?

❸ What is the difference between sperm and semen?

❹ Imagine you are a sperm cell! Describe your journey from the testes to an egg which you fertilise.

❺ A big tail helps a peacock to attract a mate, but can you think of any disadvantages of having such a huge tail?

KEY WORDS

erection
semen
ejaculation
fertilisation
contraception
courtship

Pregnancy

- ▸▸ What does the fetus need and how does it get it?
- ▸▸ How is the fetus protected?
- ▸▸ How does the fetus develop?
- ▸▸ What does the mother need to do to care for the fetus?

By 14 weeks it has all of its body organs and is called a 'fetus'

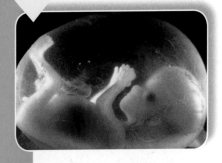

By 5 months the fetus looks like a baby, but it is too small to survive and still has a lot of growing to do!

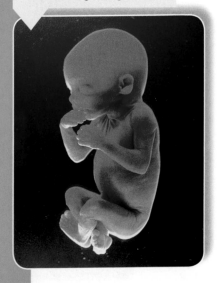

Making a baby

When an egg has been fertilised it divides to make two cells. These divide again to make four cells, then eight and so on. As it divides, it keeps moving down the oviduct. When it gets to the uterus it is a ball of cells called an **embryo**. It keeps on growing and developing and after about ten weeks it has all of the main body organs. It is then called a **fetus**.

A human embryo at five weeks is about 1 cm long

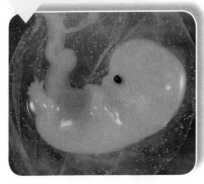

You're very lucky. You're going to have a little brother or sister!

I'd rather have a puppy.

The lining of the uterus has a rich supply of blood vessels. The embryo implants into this lining, burying itself between the blood vessels. It gets food and oxygen from its mother's blood. As the embryo grows, the **placenta** develops, joined to the lining of the uterus. Substances, like drugs and alcohol, can pass between the fetus's and the mother's blood. The mother's blood and the fetus's blood never actually mix.

The placenta

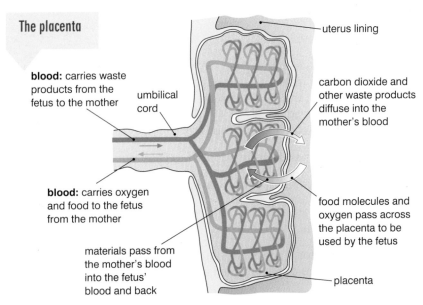

uterus lining

blood: carries waste products from the fetus to the mother

umbilical cord

carbon dioxide and other waste products diffuse into the mother's blood

blood: carries oxygen and food to the fetus from the mother

materials pass from the mother's blood into the fetus' blood and back

food molecules and oxygen pass across the placenta to be used by the fetus

placenta

- Food, water and oxygen pass from the mother's blood to the fetus's blood.
- Waste (like carbon dioxide) passes from the fetus's blood to the mother's blood.

The **umbilical cord** carries blood between the fetus and the placenta.

The fetus is inside a membrane bag, called the **amnion**. This has a liquid called 'amniotic fluid' inside it. It cushions the fetus and protects it from bumps and also helps to keep the temperature constant.

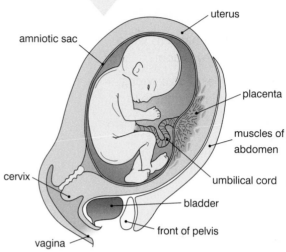

Inside the uterus

uterus
amniotic sac
placenta
muscles of abdomen
cervix
umbilical cord
bladder
front of pelvis
vagina

A healthy diet

The fetus needs the right foods so that it can develop properly:

- Calcium for healthy bones
- Iron to make blood
- Protein for growth

The fetus gets its food from its mother's blood. She needs to make sure her diet contains the right amount of these nutrients. The mother will also need extra energy from her food.

> **a** Why do you think the mother needs extra energy from her food? What food type would give her energy?

Keep out!

Some harmful things can pass through the placenta:

- **Alcohol** can cause brain damage in the fetus.
- **Carbon monoxide** from cigarettes reduces the amount of oxygen that the blood can carry. Pregnant women who smoke are more likely to have babies that are small or born too soon (**premature**).
- **Drugs**, even medical drugs, can harm the fetus. Doctors have to be careful about which drugs they give to pregnant women.
- **Viruses** cause diseases. Rubella is a minor disease in most people. In a fetus it can cause blindness, deafness and deformity.

Stretch Yourself

Make a leaflet advising a pregnant woman about taking care of herself and her unborn baby. Include information about diet and things that could harm the fetus.

Did You Know?

In the 1950s and 1960s pregnant women were given a drug called 'thalidomide'. It was given to stop them feeling sick, but it caused many babies to be born with badly deformed arms and legs.

Summary Questions

1 What is the job of: a) the placenta, b) the umbilical cord, c) the amniotic fluid?

2 Name some useful things that get through the placenta into the fetus's blood.

3 Name some harmful things that can get through the placenta into the fetus's blood.

KEY WORDS

embryo
fetus
placenta
umbilical cord
amnion
premature

Birth

▶▶ What happens when a baby is born?

▶▶ What can go wrong during childbirth?

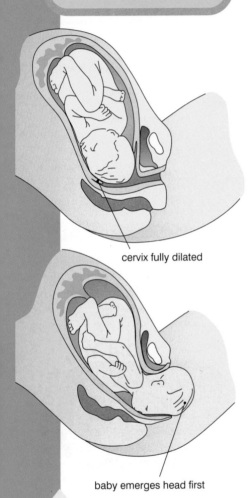

cervix fully dilated

baby emerges head first

A baby being born

Baby on the way

After about 9 months of pregnancy, the fetus is fully developed. It is ready to be born. It should have turned in the uterus so that its head is next to the cervix and it is lying upside-down.

a Look at the diagram. Why do you think pregnant women often suffer from back pain?

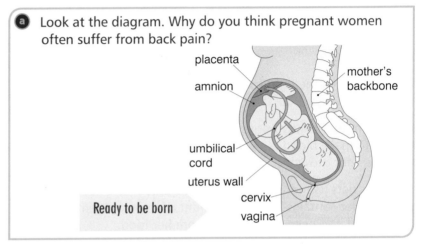

placenta

mother's backbone

amnion

umbilical cord

uterus wall

cervix

vagina

Ready to be born

Special chemicals called 'hormones' carry messages to the muscles in the uterus wall. The muscles start to contract. At first the **contractions** are weak but gradually they get stronger and happen more often. Meanwhile the muscles around the cervix relax. The cervix slowly opens.

The amnion breaks and the amniotic fluid rushes out of the vagina. This is known as 'the waters breaking'. When the cervix has opened to 10 cm, the contractions push the baby out, head first.

b What happens to the muscles in the uterus wall during birth?

c What happens to the muscles in the cervix during birth?

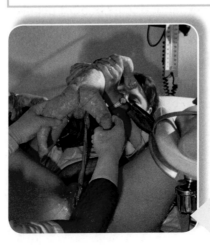

A baby is born

The baby is still attached to the placenta by the umbilical cord. The umbilical cord is cut. This doesn't hurt, as it has no nerves.

About half an hour after the baby is born the placenta is pushed out of the vagina. This is called the **afterbirth**.

Twins

Sometimes a woman might release two eggs at the same time. If they both get fertilised then twins will be produced. These will be **non-identical twins** and could be two boys, two girls or a girl and a boy.

Sometimes an egg gets fertilised then it splits and forms two fetuses. They are **identical twins** because they come from the same egg and sperm. Identical twins are always two boys or two girls. They will always look alike, as their genes are the same.

Birth complications

Sometimes a birth does not go smoothly:

- **Caesarean section** – this is where the doctor cuts through the mother's abdomen and uterus to get the baby out. There are many different reasons for doing a Caesarean section. These include the umbilical cord being squashed so the baby's blood flow is reduced. Or perhaps the cord is wrapped round the baby's neck.

- **Breech birth** – this is when the baby is not head down so it is born bottom first. Sometimes doctors do an emergency Caesarean section if the baby is in the wrong position.

- **Premature birth** – sometimes a baby is born long before the time it is due. Premature babies do not have very much body fat so they have difficulty in keeping warm. Their lungs are not fully developed. An incubator helps to keep the baby warm and provides a higher than normal level of oxygen. Sometimes their digestive system has not developed enough, so the baby is given nutrients directly into its blood. Another problem with premature babies is that their liver is not fully developed. This causes jaundice, which is treated using ultraviolet light.

The incubator keeps the baby warm and gives it extra oxygen

Summary Questions

1. Draw a flow chart to explain the stages of childbirth.

2. What is the 'afterbirth'?

3. What is the difference between identical and non-identical twins?

4. What are the problems faced by premature babies? How are premature babies looked after to overcome these problems?

KEY WORDS

contraction
afterbirth
Caesarean section
breech birth
premature birth

Growing up

- ▸▸ What do new babies need?
- ▸▸ How do humans and other animals grow?
- ▸▸ What changes take place at puberty and how are they controlled?

Great Debates

Discuss advantages and disadvantages of feeding babies with breast milk or formula milk.

Breast is best

Baby's needs

New babies have physical needs, such as food and warmth. They also have emotional needs, such as being loved and being happy.

For the first few months babies are fed on milk. Some babies are fed on formula milk. This is made from dried cow's milk with extra nutrients added to make it more like human milk. Other babies are fed on milk which is made in their mother's breasts or **mammary glands**. Milk contains the nutrients that babies need including:

- **protein** for growth
- **calcium** for making teeth and bones
- **iron** for making blood
- **fat** and **carbohydrate** for energy.

In the first few days after giving birth, women make a special kind of milk which contains antibodies.

After a few months the baby's digestive system has developed. Then it can start to eat solid foods.

> **a** What nutrient is needed for growth?

> **b** What do antibodies do?

Growth and development

New babies grow very quickly. Their cells divide to make new cells. The new cells grow then divide again.

In the first months of life, babies develop very quickly. They soon start to recognise the people close to them and can interact with them. They smile, touch and make their first attempts at 'talking'. If babies want something, they have an excellent way of getting attention – they cry!

Crying usually gets the baby the attention it wants!

Puberty

Puberty is the time when **physical changes** happen to your body. Your sex organs develop fully. You will get taller and broader at puberty. Puberty starts some time between the ages of 10 and 15. It normally lasts about two or three years. Girls usually start puberty an average of two years before boys.

Puberty is controlled by chemicals called **sex hormones**. Boys' sex hormones are made in their testes. Girls' sex hormones are made in their ovaries.

You will probably also undergo emotional changes at puberty. You will become more interested in the opposite sex. You may even find that you are more interested in the same sex. You might become more moody or find that you fall out with friends and family.

During puberty you may get spots. You will probably find that you develop a stronger body smell. You may well need to wash more often!

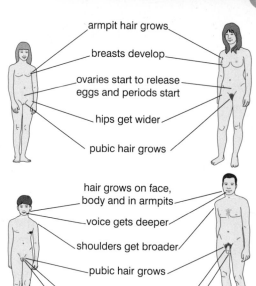

armpit hair grows
breasts develop
ovaries start to release eggs and periods start
hips get wider
pubic hair grows

hair grows on face, body and in armpits
voice gets deeper
shoulders get broader
pubic hair grows
penis and testes get bigger
testes start making sperm

From a child to an adult

Summary Questions

❶ Give three advantages of breast feeding.

❷ Use the information in the table to draw a graph to show changes in height with age.

Age (years)	0	1	2	3	4	5	6	7	8	9	10	11	12	13	14	15	16	17	18
Height (cm)	50	70	86	96	104	111	116	121	126	131	136	141	145	155	165	172	174	176	177

❸ Make a table like the one below to summarise the changes that happen at puberty.

Changes that happen to females	Changes that happen to both sexes	Changes that happen to males

KEY WORDS

mammary glands
puberty
physical changes
sex hormones

Periods

How does a woman's body control when she makes eggs?

Why do women have periods?

Time of the month

Periods are the most noticeable part of the **menstrual cycle**. This is a series of changes that happen in the ovaries and uterus. The menstrual cycle lasts about 28 days. It is controlled by special chemicals called **hormones**. The purpose of the menstrual cycle is to prepare a woman's body for pregnancy.

The lining of the uterus is made of tiny blood vessels. The lining breaks down every month if a woman is not pregnant. It comes out through the vagina with some blood. This is a period, or **menstruation**, which lasts about 4 to 7 days.

a How long does the menstrual cycle usually last?

When a period is finished, a new egg matures in one of the ovaries. The lining of the uterus starts to thicken again. About 14 days after the start of the period, the egg is released into the oviduct. This is called **ovulation**.

b What does ovulation mean?

The uterus lining continues to thicken for about a week after ovulation. The egg gets pushed along the oviduct. If it meets some sperm, it might get fertilised. A fertilised egg will start to divide, making new cells. As the cells divide they form an embryo. It passes into the uterus and implants into the lining of the uterus. The blood vessels in the lining carry food and oxygen to the embryo. They carry carbon dioxide and other waste away from the embryo.

The diagram below shows what happens during the whole menstrual cycle.

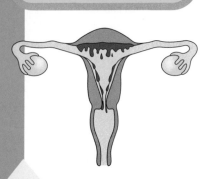

Menstruation

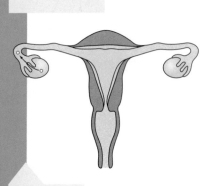

Ovulation

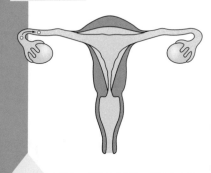

The egg moves along the oviduct

The menstrual cycle

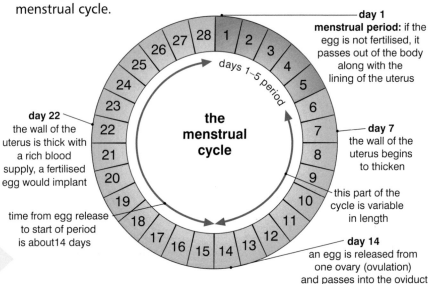

day 1
menstrual period: if the egg is not fertilised, it passes out of the body along with the lining of the uterus

days 1–5 period

day 7
the wall of the uterus begins to thicken

this part of the cycle is variable in length

day 14
an egg is released from one ovary (ovulation) and passes into the oviduct

day 22
the wall of the uterus is thick with a rich blood supply, a fertilised egg would implant

time from egg release to start of period is about14 days

the menstrual cycle

If a woman does get pregnant, the uterus lining continues to thicken. Her periods will stop until after the baby is born.

If the woman does not get pregnant, the uterus lining breaks down again. This happens about 14 days after ovulation and produces the period or bleed.

Sanitary pads or tampons are used to absorb the blood produced in periods. Periods usually occur every 28 days but this can vary a lot, especially when a girl first starts having them.

The menstrual cycle involves changes in hormone levels in the body. This can make girls and women feel tense and irritable at certain stages of the menstrual cycle. It is called pre-menstrual tension (PMT).

Tampons and sanitary pads

Stretch Yourself

What are 'hormones' and how do they control a woman's menstrual cycle?

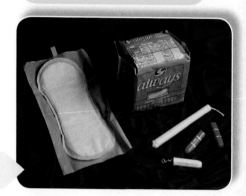

activity

Predicting periods

Susan's periods are normally every 28 days. She always makes a note of the day she starts her period. Here is a page from her calendar:

September

Sun	Mon	Tue	Wed	Thu	Fri	Sat
					1	2
3	4	5	6	7	8	Period started
10						
17						
24						

October

Sun	Mon	Tue	Wed	Thu	Fri	Sat
1	2	3	4	5	6	Period started
						14
						21
						28

November

Sun	Mon	Tue	Wed	Thu	Fri	Sat
			1	2	3	4
5	6	7	8	9	10	11
12	13	14	15	16	17	18
19	20	21	22	23	24	25
26	27	28	29	30		

● Predict what day Susan's period will start in December.

● Sperm can live for about three days after they are released. Eggs can only be fertilised for about 24 hours after ovulation. On which days in September could Susan have sexual intercourse resulting in an egg being fertilised?

Summary Questions

❶ What is probably the first sign that tells a woman she is pregnant?

❷ What type of substance controls the menstrual cycle?

❸ Why does the uterus lining thicken every month?

❹ Why do girls need more iron in their diet than boys?

❺ Why don't women have periods when they are pregnant?

KEY WORDS

periods
menstrual cycle
hormone
menstruation
ovulation

In Control

> ▶▶ How can women avoid getting pregnant if they don't want to?
>
> ▶▶ How can women be helped if they can't have a baby?

Contraception

Most people like to decide whether or not to start a family. They want to be able to have sexual intercourse without the woman getting pregnant. They do this by using **contraception**. A contraceptive is something that stops an egg being fertilised. There are several different methods of contraception.

Barrier methods

These stop the sperm from reaching the egg.

Male **condom** – this is made from thin rubber and it fits over the man's penis. It catches the sperm in a space at the end. Condoms help to stop the spread of diseases which can be passed on in sex. These are called **sexually transmitted diseases** or STDs (also called STIs – sexually transmitted infections).

Female condom – this is similar to a male condom. It fits inside the woman's vagina and catches sperm.

Diaphragm – this is a rubber disc which goes inside a woman's vagina. It stops sperm from going into her uterus. The woman has to leave it in place for a few hours after having sex. The woman has to cover it with special cream which kills sperm before she puts it inside her vagina. It must be internally fitted by a doctor.

A male condom

Chemical methods

Contraceptive pill – when a woman is pregnant, her ovaries stop releasing eggs. The 'pill' contains hormones which 'trick' the body into behaving as if the woman is pregnant, so she doesn't produce any eggs. The woman has to take the pill every day for three weeks. Then she has a week without the pill, before taking it again for three weeks. If she forgets to take it she will have to use another method of contraception. The pill provides no protection against STIs.

A female diaphragm

A contraceptive pill

Contraceptive injection – this is very similar to the pill but it is given as an injection which lasts for about three months.

Contraceptive implant – this is about the size of a matchstick. It is inserted under the skin of the arm. It releases hormones that stop the woman from releasing eggs.

Other methods

IUD or coil – this is a specially shaped device made of plastic and copper. It is inserted in a woman's uterus by a doctor. It stops fertilised eggs from implanting in the uterus lining.

No method of contraception is 100% reliable. The only way to avoid getting pregnant is by not having sex.

Link up to...

CITIZENSHIP

... ask your social, personal and health education teacher for more information about contraception.

Helping conception

Some women want to have a baby but find that they are not able to conceive (make a fertilised egg). This could be due to a problem with the man or woman. Both of them will have to go to a fertility clinic where they are tested to find out what is wrong.

Sometimes the woman's oviduct is blocked, so sperm cannot reach the egg. Sometimes men do not make enough sperm, or they make sperm that cannot swim very well. In both of these cases, IVF treatment (*in vitro* fertilisation) could help.

The woman is given hormones which make her produce lots of eggs at once. These are collected and put into a dish. Some of the man's sperm is added and eggs are fertilised in the dish. Some of the fertilised eggs are then implanted in the woman's uterus where they develop as normal.

Doctors can also collect one sperm cell and inject it into an egg using a very thin needle!

Babies born by IVF treatment are sometimes called 'test tube babies'.

Did You Know?

In 1978, Louise Brown was the first test tube baby in the world. Now she has a baby of her own, born without the need for IVF treatment.

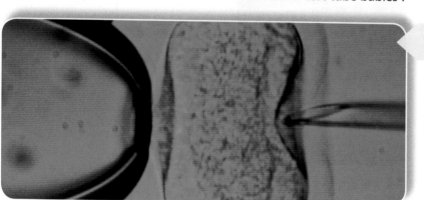

A sperm cell is injected into an egg

Summary Questions

1. Which methods of contraception help to stop the spread of sexually transmitted diseases?

2. Which are the barrier methods of contraception?

3. Draw a table to summarise the different methods of contraception and how they work.

KEY WORDS

contraception
condom
sexually transmitted disease
diaphragm
contraceptive pill
IUD

Reproduction in Plants

▶▶ What is asexual reproduction?

▶▶ What are the advantages of asexual reproduction?

▶▶ What are the disadvantages of asexual reproduction?

Daffodils reproduce asexually by making bulbs

Asexual reproduction

Plants make gametes which are used for sexual reproduction, the same as animals. But plants can also reproduce without sex, unlike most animals. This is called **asexual reproduction**. Plants reproduce asexually in a number of different ways.

Bulbs – many plants, like daffodils, make bulbs. The bulb contains food stores. The plant grows in the spring. Its leaves make food by photosynthesis. Some of the food is sent to the bulb where it is stored. When the bulbs are big enough, they split into several smaller bulbs. Next year there will be more daffodil plants.

Runners – some plants grow runners. These are stems which grow sideways from the plant. A small plantlet starts to grow at the end of the runner. When the plantlet touches the ground it forms roots. Eventually the runner disappears, leaving the new plant. Spider plants and strawberries reproduce using runners.

Spider plants make runners

A potato plant

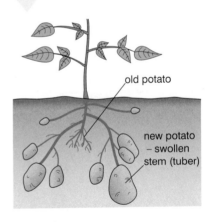

old potato

new potato – swollen stem (tuber)

Tubers – a potato is a tuber. Potato plants store food underground in the potatoes. Each potato plant makes many potatoes. Next year a new plant grows from each potato.

Asexual reproduction is very useful for the plant. It makes new plants quickly and easily. Each plant is exactly the same as the parent. This means it should fit in really well where it is growing.

There are disadvantages to asexual reproduction:

- If conditions change the plants might not fit in so well.
- If they are all identical they might all die.
- If a disease kills one plant it will probably kill them all.

> **a** Give one advantage and one disadvantage of asexual reproduction in plants.

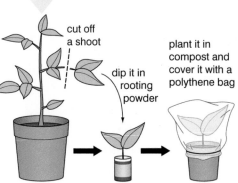

Taking a cutting

cut off a shoot

dip it in rooting powder

plant it in compost and cover it with a polythene bag

Humans make use of asexual reproduction in plants. Imagine a strawberry grower has a strawberry with perfect taste and colour. He will want to grow lots of strawberries which are exactly the same. Growing strawberries from runners makes sure they are the same.

Other crops like potatoes are produced by asexual reproduction, so that we always get the same varieties of potatoes.

Humans also use artificial asexual reproduction in plants, for example, by taking cuttings.

activity

Cuttings

- Take a geranium plant.
- Cut a stem off just below where a leaf joins.
- Remove the lower leaves so there are two or three leaves on the plant.
- Dip the cut end in rooting powder. This helps the stem to make roots. Avoid breathing dust.
- Put the cutting in compost.
- Cover the pot with a polythene bag. This keeps the plant moist. After two or three weeks it will grow roots.

We use cuttings to produce many food plants, such as apples. There are many varieties of apple. To get the variety of apple you want, you should grow them from cuttings taken from a parent plant.

Different varieties of apples

Summary Questions

1 What is a tuber?

2 Draw a table to show the advantages and disadvantages of asexual reproduction for plants.

3 Draw a table to show the advantages and disadvantages of asexual reproduction in plants for gardeners and farmers.

KEY WORDS

asexual reproduction
bulb
runner
tuber

Flowers and Pollination

▸▸ How do plants reproduce sexually?

▸▸ How does pollen get from one flower to another?

Flowers

Plants can also reproduce sexually. They make male sex cells called **pollen**. They also make female sex cells called **ovules**. The male and female sex organs are both part of the flower. Remember that the next time you give someone a bunch of flowers! The male parts are called **stamens**. The female parts are called **carpels**.

A typical flower

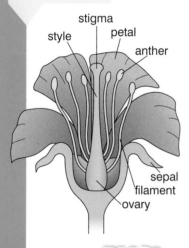

stigma
style
petal
anther
sepal
filament
ovary

activity

Looking at flowers

Your teacher will give you a flower to look at.

● Try to find the parts which are labelled in the diagram. (This can be a bit tricky because all flowers are different.)

● As you identify each part, remove it carefully with tweezers.

● On a piece of paper, arrange the parts the same as they were in the flower but spread out so they are separate. Use sticky tape to hold them in place.

● Label each part.

In the male part of the flower, the stamen:

● Pollen is made in the **anthers**.
● **Filaments** support the anthers.

In the female part, the carpel:

● The ovules are made in the **ovary**.
● The **stigma** and **style** are above the ovary.

For fertilisation to happen the male sex cell and the female sex cell have to join. This is easy enough in animals, as they can move around. But how do plants manage it?

Did You Know?

The biggest flowers in the world are from a plant called rafflesia. The flower can be over 1 m across. It looks and smells like rotting meat! What kind of insect pollinates it?

ⓐ Name the male parts of a flower.

ⓑ Name the female parts of a flower.

Pollen carried by wind and insects

Pollination

Pollination is the transfer of pollen from the anthers to the stigma. There are two main ways this happens:

● by insects
● by wind

Insect pollination

Insects, such as bees, butterflies and moths, carry pollen from plant to plant. But why would they do that? Flowers make a sweet, sugary liquid called nectar. Insects visit flowers to drink nectar. Plants need to make sure the insects can find them, so they 'advertise'! They make flowers with big, brightly coloured petals which insects can see. They make nice smells which insects can detect.

When a bee visits a flower, some pollen brushes from the anthers onto its hairy body and sticks to it. The pollen is quite sticky. When the bee lands on another flower some of the pollen brushes off onto the sticky stigma.

Insects are important carriers of pollen, but birds, bats and even reptiles also carry pollen from flower to flower.

> **c** Why do insects go to flowers?

> **d** Why do you think that flowers that are pollinated by moths have a particularly strong smell?

Wind pollination

Not all plants are pollinated by insects. Many, such as grass, are pollinated by the wind. You may not have realised that grass has flowers. Since they do not need to attract insects, they do not have large bright petals. Wind-pollinated flowers have anthers that hang out from the flowers. This means that the pollen is easily picked up by the wind. The pollen is very light.

These plants make lots of pollen because most of it doesn't reach the stigmas. The stigmas also hang from the flowers so the pollen is caught as it blows on the wind. The stigmas often have lots of branches that form a net to catch the pollen.

The anthers and stigmas can be seen clearly on this grass flower

Summary Questions

1 Copy and complete:

The male parts of a flower are called the _____. The _____ makes pollen. It is supported by the _____.the female parts are called the _____, style and stigma. _____ are made in the ovaries.

2 Make a table to show the differences between insect-pollinated and wind-pollinated flowers. Explain why there are differences between the two types of flower.

3 Hay fever is caused by pollen. Do you think it is more likely to be from wind- or insect-pollinated flowers? Explain your answer.

KEY WORDS

pollen
ovule
stamen
carpel
anther
filament
ovary
stigma
style

Fertilisation in Plants

▸▸ How does fertilisation happen in flowers?

▸▸ What factors affect the growth of pollen tubes?

A carpel

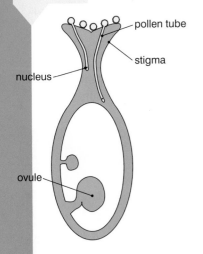

pollen tube

stigma

nucleus

ovule

Pollination gets the pollen from the anther to the stigma, but that is only part of what has to happen. From the stigma to the ovary is still a long way.

a What are plant's male sex cells called?

b What are plant's female sex cells called?

Pollen lands on the stigma, but its nucleus has to fuse with the ovule nucleus. How can it do this?

activity

Growing pollen tubes

- Shake a flower so that some pollen falls on a microscope slide.
- Add a few drops of sugar solution.
- Make another slide but add water instead of sugar solution.
- Put a cover slip on top of each slide and leave it for about 20 minutes.
- Use a microscope to look at the slides.

You will see pollen tubes growing in the sugar solution.

A sticky sugar solution on the top of the stigma stops pollen from falling off or blowing away. It also makes **pollen tubes** grow from the pollen. These tubes grow down the style and into the ovary. When it reaches the ovule, the pollen nucleus moves down the tube. It fuses with the ovule nucleus. The ovule has been fertilised.

c Why is the top of the stigma sticky?

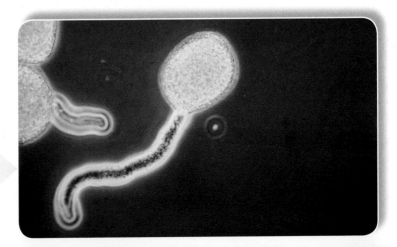

A pollen tube growing from a pollen grain

The fertilised ovule now develops into a seed. Most parts of the flower die and wither away. The ovary forms a fruit.

Fertilisation

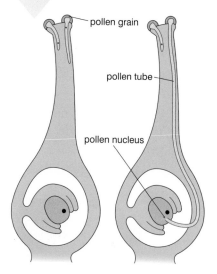

pollen grain

pollen tube

pollen nucleus

What conditions are best for growing pollen tubes?

Your teacher will give you five different concentrations of sugar solution.

Your task is to find out which concentration is best for growing pollen tubes.

- What is the **independent variable** in this investigation? (See page 156.)
- What is the **dependent variable** in this investigation?
- What variables will you need to control?
- How will you record your results?
- How will you display your results?
- How will you know which concentration was the best for growing pollen tubes?

Summary Questions

❶ How does the pollen nucleus reach the ovule nucleus?

❷ What is the difference between pollination and fertilisation?

❸ Write out these sentences in the correct order:

A pollen tube grows from the pollen grain.

The pollen nucleus fuses with the ovule nucleus.

The fertilised ovule develops into a seed.

A pollen grain lands on the stigma.

The pollen nucleus moves down the pollen tube.

❹ Copy and complete the table to summarise the differences between sexual reproduction in animals and plants:

	Reproduction in plants	Reproduction in animals
Name of male sex cell		sperm
Name of female sex cell		
Where are male sex cells made?	anther	
Where are female sex cells made?		ovaries
How does the male sex cell reach the female sex cell?		

KEY WORDS

pollen tube
independent variable
dependent variable

Spreading the Seeds

How are seeds dispersed from the parent plant?

What conditions do seeds need in order to germinate?

Burdock fruits have stuck to this puppy

Dispersal

In everyday life a **fruit** is something nice to eat. To a scientist a fruit is something that helps in the **dispersal** of seeds. That means spreading the seeds away from the parent. In this way the seeds have a better chance of surviving. They won't be in the shade of the parent. They won't have to compete with the parent plant for water and minerals from the soil. There are four main ways this happens:

● by animals
● by wind
● by water
● by self

Animal dispersal

The fruits we eat help seeds to disperse. When you eat a tomato you eat the seeds as well. The seeds have a hard coat so they are not digested. They pass through your body in your faeces (poo). If you are ever lucky enough to visit a sewage works you may see tomato plants growing there!

When animals eat fruit, the seeds have a nice pile of manure to start growing in.

Some plants make fruits with hooks on them. These catch on animals' fur. They can be carried a long way before they fall off.

Wind dispersal

Some seeds are spread by the wind.

Each dandelion 'parachute' is a fruit. They are light and carry the seeds on the wind.

Some trees such as sycamore, ash and elm make fruits with wings. When they fall from the tree, they spin like helicopters. They are carried further by the wind.

Water dispersal

The coconut is one of the biggest seeds. Coconut palms grow near the sea. The seeds fall into the sea and can float for thousands of miles before being washed onto the beach. Coconuts are covered in fibre, which traps air so they can float. This is removed before they get to the shops.

Self dispersal

Some fruits are designed to explode as they dry out. This sends the seeds flying through air, away from the parent!

Exploding witch hazel fruits

activity

Helicopter fruits

Investigate how the height a 'helicopter' type seed falls from affects the time it takes to reach the ground.

- What is the independent variable? (See page 156.)
- What is the dependent variable?
- What variables will you have to control?
- How can you make your investigation more reliable?

Germination

When seeds start to grow this is called **germination**. How do seeds know when to germinate? Why don't they germinate in the packets before you buy them? What conditions do seeds need to germinate?

activity

Investigating germination

Set up the experiment shown below.

Water normally has oxygen dissolved in it. Boiling the water removes the oxygen.

Leave the investigation for a week.

Copy and fill in the results table:

Tube	Does it have oxygen?	Does it have water?	Does it have warmth?	Did the seeds germinate?
A				
B				
C				
D				

- What conditions were needed for the seeds to germinate in the activity?

Summary Questions

1. Why is it better for seeds to grow well away from the parent plant?
2. Make a poster to show the different ways that seeds are dispersed.
3. Explain why each of the conditions needed for germination is important.

KEY WORDS

fruit
dispersal
germination

know your stuff

▼ Question 1 (level 3)

The diagrams below show a human egg and sperm.

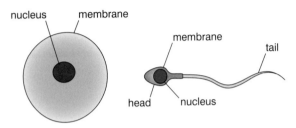

Human ovum (egg)   **Human sperm**

a What are eggs and sperm?
Choose from the list below:

   animals   cells   organs   [1]

b Which part of a sperm cell is used to help it swim? [1]

c Where are sperm cells made? [1]

d (i) What is the function (job) of the nucleus of a sperm? [1]

  (ii) Sperm use their tails to swim to the egg so that fertilisation can take place. Give one other way in which a sperm is adapted for fertilisation. Explain your answer. [1]

e The diagram below shows a sperm fusing with an egg.

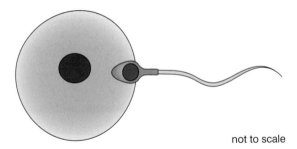

not to scale

What is the name of this process?
Choose from the list below:

fertilisation   photosynthesis

pollination   respiration   [1]

▼ Question 2 (level 4)

The diagram shows a section through the female reproductive system.

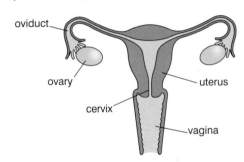

a (i) How often are eggs normally released in the female reproductive system? [1]

  (ii) Where does fertilisation normally take place? [1]

  (iii) Where does the fetus develop? [1]

b A woman has blocked oviducts. How would this stop her from getting pregnant? [1]

c What are the missing words in the sentences below?

A fertilised egg divides into a tiny ball of cells called an embryo.

The embryo attaches to the lining of the uterus. Here the embryo grows to become an unborn baby, called a i) ....

It takes about ii) .... months for a baby to develop inside its mother. [2]

d The diagram below shows a baby growing in its mother's uterus.

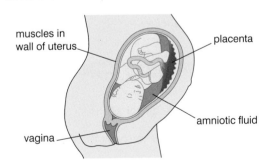

know your stuff

(i) What is the function (job) of the amniotic fluid? [1]

(ii) Some harmful substances, such as nicotine, can pass from the mother's blood to the baby's blood. Through which part does this happen? [1]

e Substances can pass from the mother's blood to the fetus's blood. Other substances can pass from the fetus's blood to the mother's blood.

Which way, if any, do the substances in the table pass? Copy the table and tick one box in each row.

Substance	Passes from the mother's blood to the fetus's blood	Passes from the fetus's blood to the mother's blood	Does not pass between the mother's blood and the fetus's blood
Poisons from cigarette smoke			
Oxygen			
Digested food			
Carbon dioxide			

[4]

f Name one useful substance, other than food and oxygen, which passes from the mother to the fetus. [1]

g How does the wall of the uterus push the baby out when it is born? [1]

▼ **Question 3 (level 6)**

Use words from the list to copy and complete the sentences about the menstrual cycle. [4]

daily   uterus   middle   end
an ovary   weekly
beginning   monthly   vagina

Menstruation is part of a **a**.................... cycle.

The cycle begins when the lining of the **b**.......................... breaks away.

An ovum (egg) is released from **c**.................at about the **d** .... of each cycle.

▼ **Question 4 (level 7)**

a The diagram shows the male reproductive system.

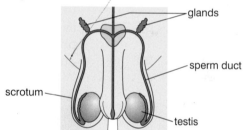

A man's testes are outside the main part of the body. What is the relationship (link) between sperm production and temperature? [1]

b Chemicals, called sex hormones, cause changes in boys' and girls' bodies. They start to be made when people reach puberty.

(i) Where are sex hormones produced in boys? [1]

(ii) Describe two of the changes in boys' bodies and two of the changes in girls' bodies. [4]

c (i) About 200 million sperm are released into the body of a woman during sexual intercourse. Why is it important to release so many sperm? [1]

(ii) Sperm cells are mixed with a fluid to make semen. Semen contains sugar. Explain how and why a sperm makes use of the sugar. [2]

Reversible and Irreversible Changes

When bread is cooking you observe a smell and see that the dough is rising

What is happening around us?

- When you go into a bakery you can tell if a fresh batch of bread has just been baked. We use our senses. You can smell the bread and can hear the pots and pans clattering. You can see that the crust has browned and feel that the bread is warm. You can even taste it! These are all observations. In science we use our senses to make observations.

a What are our five senses?

b Why should you never use your sense of taste in the lab?

- When you observe an experiment in a science lab, it is important to record your findings. Then other people can use your work to help them as well. This is why you should record your observations in a clear way. In most experiments we record our observations in the form of a table.

Making observations

- Put some soap onto beetroot.
- Now add some lemon juice.
- What do you observe?

You will notice that the beetroot changes colour.

If you add the soap again, it will change back. This is an example of a reversible chemical reaction.

- To be a scientist you should observe the world around you. In a lab it is important that you use your eyes, ears, skin and nose. You should always take care in a lab to prevent any accidents.

⚠ **Safety:** Most of the time, you will need to wear eye protection to stop things splashing into your eyes. When you observe using your sense of smell, use your hand to waft the smell up your nose. Do not take a big breath in.

⚠ **Safety:** To observe with your skin, gently touch the outside of the container to feel if it is hot or cold.

c Why should you never take a big sniff of an unknown chemical?

d When you are doing an experiment, how can you make sure that you are safe?

When you are observing using touch, gently touch the outside of a container. NEVER put your hands into chemicals

This is how to do a scientific sniff, dear.

When you are observing a smell, waft the smell up your nose. NEVER take a big sniff!

activity

Observing chemicals

Your teacher has set up six different chemicals, labelled A to F. You are going to mix two chemicals together in one test tube. Use your senses to make observations. You will need to record your results in a table. Copy and complete the table below:

Chemicals	What I observed by looking	What I observed by smelling	What I observed by hearing	What I observed by touching
A+B				
C+D				
E+F				

Carefully mixing the chemicals

What are Reversible and Irreversible Changes?

▶▶ What is a 'reversible' change?

▶▶ What is an 'irreversible' change?

▶▶ How do we know what type of change is happening?

All change here

Changes happen all around us, all the time. Some of these changes can be reversed and some cannot.

a Make a list of some changes that happened to make your breakfast.

When we want to make ice cubes, we fill a tray with water and put it in the freezer. Soon the water turns into ice. But if we leave the ice cubes out of the freezer, they will melt back into water. The substance is unchanged and can freeze, then melt, again and again. This is an example of a **reversible** change, in which no new substance is made. We call this a **physical change**.

b What do we mean by a 'physical' change?

c Give an example of an everyday physical change.

Some changes are not reversible. When we burn a match, you cannot collect the ash and gases and turn it back into the unlit match. Even if we collected everything you could not get the match back to what we started with. This is an example of an **irreversible** change. We call this type of change a **chemical change**.

d What do we mean by a 'chemical' change?

e Give an example of an everyday chemical change.

Water can freeze into ice, and ice can melt into water. This is a reversible change

You can only burn a match once

If we observe changes carefully we can decide if they are reversible or irreversible. Scientists write their observations in tables so that it is easier to see patterns in their results.

If a new substance has been made, then a chemical change has happened. New substances might be made when we see a colour change or bubbles appear. But, if the starting substance can easily be re-formed, then a reversible change has happened.

> **f** At home, Simon added some baking powder to vinegar. He observed bubbles and a fizzing noise. What type of change is this?

Summary Questions

1. Copy and complete using the words below:

   reversible   change   chemical   irreversible   physical

   When water turns into steam, it can easily be turned back. This type of … is called a … change. If no new substance is made then a reversible change is also a … change. When an egg fries, you can't get the raw egg back. This is an … change or a … change.

2. What type of change, physical or chemical, is each of the following an example of?
   a) melting chocolate
   b) burning a match
   c) freezing an ice lolly
   d) cooking meat

3. Look carefully at the cartoon called 'It's all change!', how many physical changes can you spot? How many chemical changes can you spot?

4. What are the differences between chemical and physical changes?

It's all change!

KEY WORDS

reversible
physical change
irreversible
chemical change

Are All Acids Dangerous?

Acids

Some acids can eat away (corrode) metal, while other acids are safe to eat!

Some foods like oranges contain natural acids. If acids are safe to eat, they are usually weak acids. Weak acids taste sharp or sour. Some foods and drinks have acids added to them so that they stay fresh for longer.

Some acids are safe to eat!

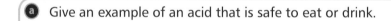

ⓐ Give an example of an acid that is safe to eat or drink.

ⓑ Give an example of an acid that you can find in the school lab.

Hydrochloric acid is a strong acid

Some acids are dangerous. We call these 'strong acids', for example hydrochloric acid. These strong acids have special **hazard** symbols on their containers to warn us of the dangers. The three hazard symbols that we might find on an acid are **harmful**, **irritant** and **corrosive**.

Explanation	Hazard symbol
Can make us very ill if we eat them, breathe them in or absorb them through our skin.	Harmful ✖ HARMFUL
Will make our skin red and itchy.	Irritant ✖ IRRITANT
Kills living cells and can eat away at a lot of different materials, such as metals.	Corrosive CORROSIVE

Next time you...

... are travelling by car, try to spot some chemical tankers and look carefully near their delivery hose and on the side of the truck for some hazard symbols. Can you remember what they mean? Why is it important that the tankers have these symbols on them?

c What hazard symbol would you put on the bottle of concentrated hydrochloric acid?

d Why are hazard symbols more useful than just writing the word on the container?

Summary Questions

1 Copy and complete the sentences by matching the two columns:

a)	Not all acids	which is a weak acid.
b)	Lemons contain citric acid,	are dangerous.
c)	Some acids are dangerous	can kill cells.
d)	Corrosive acids	and will have a hazard label on their bottle.
e)	Harmful acids can make you	ill if they get into your body.

2 Imagine that you spill hydrochloric acid onto the bench in the school lab. What should you do?

KEY WORDS

acid
hazard
harmful
irritant
corrosive

Are all Alkalis Dangerous?

Alkalis

▸▸ Which substances are alkalis?

▸▸ What are some of the properties and uses of alkalis?

Alkalis are another important group of chemicals that you can find all around you. Alkalis will undergo a chemical reaction with acids. Like some acids, you can safely touch some alkalis. They feel soapy. These chemicals are often examples of **weak** alkalis.

Some alkalis are dangerous. These are **strong** alkalis and can have hazard symbols on their bottles, just like strong acids. Do not touch these!

Did You Know?

When an acid and an alkali meet, an irreversible change happens. The chemical reaction is called neutralisation. See page 64.

ⓐ What hazard symbols do you think could be on a bottle of alkali?

You can often use alkalis as cleaning products. Soap is an example of a weak alkali, but it is still an irritant.

So where do we use strong alkalis? These are also used around the house for cleaning. Oven cleaner is a strong alkali. If we are going to use these dangerous chemicals, we must protect our skin. Sodium hydroxide is an example of a strong alkali that you could find in the school lab.

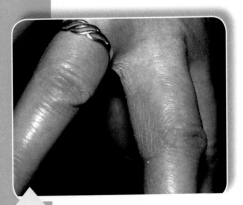

If you leave soap on your skin, it will go red and itchy. This is because soap is an irritant.

Wear gloves when using dangerous chemicals – even at home

Rubber gloves would normally be enough for most people.

▷ **Safety:** If you get a strong acid on your skin, it can cause a chemical burn. If you get a strong alkali on your skin, a chemical burn can also be caused but in a different way. It is important to wash any skin that has touched a strong acid or alkali with plenty of cold water and to tell an adult. If you get a strong alkali in your eye, you need to get medical help straight away. This is because they react with the fat in your cells and this can cause blindness.

b What should you do if you get a corrosive alkali on your skin?

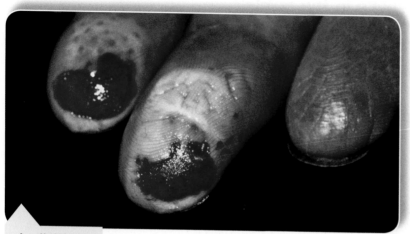

An alkali burn

Stretch Yourself

Alkalis belong to a bigger group of chemicals called bases. Bases will also react chemically with acids. An alkali is a base that can dissolve in water. All alkalis are bases but not all bases are alkalis. Can you find some examples of bases and alkalis?

Summary Questions

1 Copy the words below and match them to their correct definitions:

a) alkali	an acid or alkali that would be corrosive or harmful
b) weak	a chemical that will react with an acid
c) strong	a base that can dissolve in water
d) base	an acid or alkali that is an irritant

2 Sodium hydroxide, potassium hydroxide and calcium hydroxide are all examples of strong alkalis. Make a list of things that they have in common.

3 Have a good look around your home and make a list of alkalis and the hazard symbols that are on the labels. Do not open any containers. Record the information in a table.

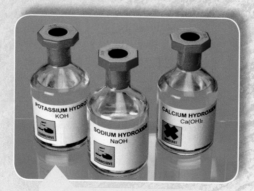

Alkalis that you may find in your school lab

KEY WORDS

alkali
weak
strong

Indicators

- ▶▶ What is an indicator?
- ▶▶ What does the pH scale show us?
- ▶▶ How do we test the pH of substances?

There's no way you could tell the difference between these three liquids without the labels... They all look the same!

What's in the bottle?

Most acids and alkalis that we use at school are colourless solutions and all look the same. We can use a special chemical called an **indicator** to tell us if a chemical is an acid or alkali.

> ## activity
> ## Making an indicator
>
> You can make you own indicator:
>
> - Just cut up some red cabbage and mash it with a little water.
> - Test the red cabbage juice by putting a little lemon juice into it. What colour does it go?
> - Add some soap to some more red cabbage juice. What colour does it go now?
>
> You have made a simple indicator.

There are lots of different indicators that we use. **Universal indicator** is one of the most popular because it is a mixture of different dyes and gives us a range of different colours. Then we can match the colour to a **pH** number. The pH number tells us how strongly or weakly, acidic or alkaline, the solution being tested is.

Indicators mainly come from plants

The pH scale

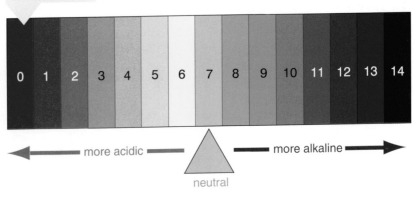

0 1 2 3 4 5 6 7 8 9 10 11 12 13 14

◀━━ more acidic ━━   ━━ more alkaline ━━▶

neutral

(a) What colour is a solution of a strong acid on the pH scale?

(b) What type of chemical would be purple with universal indicator?

Some chemicals have a pH of 7. This means that they are not acidic or alkaline. They are **neutral** chemicals. Neutral chemicals include water, blood and alcohol.

c What type of chemical would be green with universal indicator?

Link up to...

MATHS

For each decrease of one unit on the pH scale, the solution becomes 10 times more acidic! How much more acidic is a solution with a pH of 3 than one with a pH of 5?

activity

Testing the pH of solutions

At home there are lots of different chemicals, in the food we eat to the cleaning products that we use. We can find the pH number of these substances using universal indicator. In these substances, there is one chemical that is either an acid, alkali or neutral and it will have a scientific name. Lemonade is an acid, because it contains citric and carbonic acid.

You are going to test some solutions to find out how acidic or alkaline they are. Before you start your experiment it is important to draw a table to **record** your results. You are going to record the colour that the solution turns with universal indicator and its pH number. In the last column classify the solution as strongly or weakly acidic or alkaline.

Solution	Colour of indicator in solution	pH of solution	Strongly or weakly acidic or alkaline?
A			
B			
C			

- Put about 1 cm depth of a solution into a test tube.
- Add a few drops of universal indicator.
- Compare the colour of the solution with the pH chart.
- Record your results.

⚠ **Safety:** Wear eye protection!

1 Copy and complete the sentences below:
  a) Special chemicals called indicators ...
  b) Universal indicator is a mixture ...
  c) Universal indicator can be used ...
  d) Indicators can be ...

2 What properties should an indicator have?

3 We can use beetroot juice as an indicator. Why does pickled beetroot show us the colour of the indicator in acid?

Summary Questions

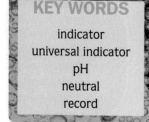

KEY WORDS

indicator
universal indicator
pH
neutral
record

Acid Reactions: Neutralisation

Mix it up

When we mix an acid and alkali together an irreversible change happens. This **chemical reaction** is called **neutralisation** and a **neutral** solution can be made.

$$acid + alkali \rightarrow metal\ salt + water$$

> What is a neutral chemical?
>
> What happens when an acid and alkali meet?
>
> How do I measure the volume of a liquid?

a What is neutralisation?

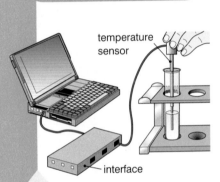

A data-logger is used to monitor pH

Farmers use neutralisation

activity

Neutralisation

The pH of waterways is very important to the fish and plants that live there. If the pH changes too much, then plants and animals could die. Scientists take samples and check the acidity of the water. If the pH is not right, they add weak acid or alkali to neutralise the extra chemical. If endangered plants or animals live in the water then the pH is monitored all the time, using a data-logger.

- Measure $10\,cm^3$ of $0.1\,M$ sodium hydroxide into a beaker.

- Add a few drops of universal indicator. What colour does the solution go? What does this tell you about the solution?

- Put in the pH probe into the solution and start recording the data.

- Add $9\,cm^3$ of $0.1\,M$ hydrochloric acid into the beaker and mix. What colour is the solution now? What does this tell you about the solution?

- Using a dropping pipette, add drops of hydrochloric acid and stir until your solution turns green. What type of solution have you made? What should you do if you add too much acid, to get back to green?

- Stop recording the data and print off your graph.

- Colour the pH line in red where the solution is acid, and blue where the solution is alkaline. Put a green cross on the pH line where the solution is neutral.

 Safety: Wear eye protection!

Neutralisation is an important chemical reaction used in everyday life. Stomach acid helps us to digest our food and it protects us from some illnesses. But, sometimes stomach acid travels up your food pipe (oesophagus) and causes a burning feeling in your chest (heartburn). We use weak alkalis in medicines to stop heartburn. You can take a weak alkali, called an **antacid,** as a tablet or liquid. The antacid neutralises the extra acid and stops the uncomfortable burning feeling.

If you have heartburn then a weak alkali can help you feel better

b What properties do you think a good antacid would have?

This will neutralise the problem!

Did You Know?

Wasp stings are alkali and these can be neutralised by an acid such as vinegar.

Next time you...

... get stung by a nettle, make sure you rub the sting with a dockleaf. Acid causes the irritation. When you rub the area with a dock leaf, the juice is an alkali and this neutralises the sting and stops it from hurting.

Summary Questions

1 Copy and complete the sentences by matching the two columns:

a)	Neutralisation is a	acidic or alkaline and have a pH of 7.
b)	Antacids are medicines	a neutral solution of a salt and water.
c)	Neutralisation makes	chemical reaction between an acid and an alkali.
d)	Neutral solutions are neither	made of weak alkalis.
e)	Antacids	neutralise extra stomach acid that can cause heartburn.

2 Copy and complete the following neutralisation word equations. The general equation in the opening paragraph will help you.

a) hydrochloric acid + sodium hydroxide → sodium chloride + ...

b) nitric acid + ... → sodium nitrate + water

c) ... + potassium hydroxide → potassium nitrate + water

3 Design an experiment to compare three different antacids, to decide which is the best one to stop heartburn. You need to think about what makes a good antacid and how you would make your investigation a fair test.

KEY WORDS

chemical reaction
neutralisation
neutral
antacid

Acid Reactions: Metals

>> What do we observe when a metal reacts with an acid?

>> How do we test for hydrogen gas?

Eating away

Corrosion is when metals react with chemicals in the environment. Rain water is actually carbonic acid with a pH of about 5. Metals used in buildings can corrode because of their reaction with rain water.

$$acid + metal \rightarrow metal\ salt + hydrogen$$

a What type of chemical is rain water?

Corrosion can cause a lot of expensive damage

Some metals, like gold, are very unreactive and do not corrode. Other metals, like calcium, are very reactive and when they meet an acid they make **hydrogen** gas.

b Steel can corrode, but gold cannot. Why are buildings made of steel rather than gold?

The Hindenburg crash

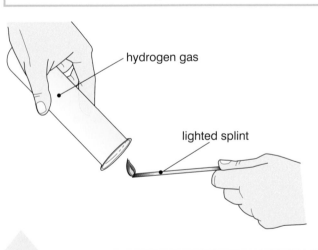

hydrogen gas

lighted splint

To test for hydrogen gas, put a lighted splint to the mouth of a test tube of the gas. If you hear a squeaky pop, hydrogen was in the tube.

Metals and acids

Metals are used in construction to hold up our roads and buildings. Acids in the environment can have a chemical reaction with some of these metals and this will change their properties. This is an example of 'chemical weathering' (you will find out more about this in Fusion 2).

You are going to list three different metals in order of **reactivity** by observing their reaction with hydrochloric acid.

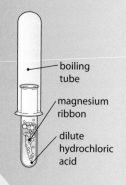

- Measure 3 cm³ of the hydrochloric acid into a test tube.
- Add a piece of metal.
- Make and record your observations.
- Put a boiling tube over the test tube and wait about 20 seconds.
- Remove the boiling tube, and quickly hold a lighted splint at its mouth.
- Make and record your observations.
- Repeat with the other metals.
- List the metals tested from the most reactive to least reactive. How did you decide on the order?
- Was your test fair? Explain your answer.

boiling tube

magnesium ribbon

dilute hydrochloric acid

Don't forget to use all of your senses except taste to observe

 Safety: Wear eye protection and keep the metals away from flames.

❶ Copy the key words below and match them to their correct definitions:

a) corrosion	when a metal has a chemical reaction
b) hydrogen	how fast a substance irreversibly changes
c) reactivity	a gas that you can test with a lighted splint and hear a squeaky pop!

Summary Questions

❷ Copy and complete the following word equations. The general equation in the opening paragraph will help you.

a) hydrochloric acid + magnesium → magnesium chloride + …

b) nitric acid + … → magnesium nitrate + hydrogen

❸ Risk assessments include all the risks in a practical, how the risk can be reduced and what to do if an accident happens. Risk assessments can be completed in a table similar to the one below:

Risk	How to reduce the risk	What to do if there is an accident
Hydrochloric acid – irritant	Wear eye protection.	Tell the teacher. Remove any clothes with the acid on. Wash skin for 10 minutes under cold water.
Glassware	Use a test tube rack.	Tell the teacher. Use a dust-pan and brush to put the pieces in the glass bin.

Write a full risk assessment for the activity above.

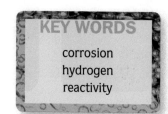

KEY WORDS

corrosion
hydrogen
reactivity

Acid Reactions: Metal Carbonates

- ▸▸ What do we observe when copper carbonate reacts with an acid?
- ▸▸ How do we test for carbon dioxide gas?

Reacting rocks

Many rocks contain metal **carbonates**. The White Cliffs of Dover are made of chalk. These cliffs mainly consist of calcium carbonate. Malachite is a green mineral and can be used in jewellery. Malachite contains mainly copper carbonate.

Metal carbonates react with acid to make **carbon dioxide** gas, water and a metal salt.

We can show the chemical reaction between a metal carbonate and an acid in a word equation:

acid + metal carbonate → metal salt + carbon dioxide + water

Many rocks contain metal carbonates

a How do you know that the reaction between a metal carbonate and acid is a chemical change?

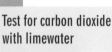
... eat a fizzy sweet, think about the chemical reaction going on. Sodium carbonate and citric acid are two solid chemicals that are safe to eat. They are used to make fizzing sherbet. When you put it in your mouth, the spit (saliva) lets the two chemical mix, making a gas and giving you a tingly feeling.

Carbon dioxide makes carbonated drinks fizzy

Carbon dioxide gas is denser than air and it sinks. You can test for carbon dioxide by bubbling it through **limewater**. If this alkaline solution turns cloudy, the gas is carbon dioxide.

b What is the test for carbon dioxide?

Test for carbon dioxide with limewater

Metal carbonates and acids

activity

You are going to investigate the reaction between copper carbonate and hydrochloric acid.

- Measure 3 cm³ of the hydrochloric acid into a test tube.
- Add half a spatula of copper carbonate (HARMFUL).
- Quickly put on the bung with the delivery tube going into a test tube containing limewater (IRRITANT).
- Make and record your observations.

 Safety: Wear eye protection.

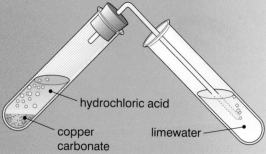

hydrochloric acid

copper carbonate

limewater

Summary Questions

❶ Copy and complete the sentences below:
   a) Metal carbonates can be found in ...
   b) Metal carbonates react ...
   c) Carbon dioxide, a metal salt and ...
   d) Carbon dioxide gas can be tested ...

❷ Copy and complete the following word equations. The general equation in the opening paragraph will help you.
   a) hydrochloric acid + calcium carbonate → calcium carbonate + ... + water
   b) nitric acid + calcium carbonate → calcium nitrate + carbon dioxide + ...
   c) ... + sodium carbonate → sodium chloride + carbon dioxide + water

❸ When calcium carbonate pieces are put into hydrochloric acid or nitric acid they react, but when they are put into sulfuric acid they react at the start, but the reaction quickly stops. Complete some research to explain this observation.

Link up to...

GEOGRAPHY
Minerals are found in rocks. When there is enough metal in the rock and it is worth extracting, the rock is called an **ore**. You will find out more about carbonate rocks and minerals in your geography lessons.

KEY WORDS
carbonate
carbon dioxide
limewater

Is it a Metal?

- ▶▶ What is a metal?
- ▶▶ What is a non-metal?
- ▶▶ How do we know if a substance is a metal or non-metal?

Classifying materials

We group most materials as either **metals** or **non-metals**. Each group has, more or less, opposite **properties**. Metals are shiny, **malleable** (bend easily) and are mostly solid at room temperature. Non-metals can look dull and are **brittle** (break easily) if they are solid.

General properties of metals

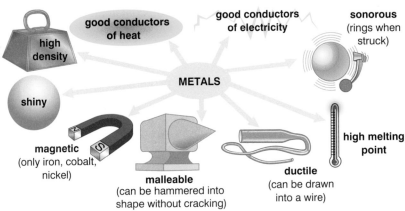

high density

good conductors of heat

good conductors of electricity

sonorous (rings when struck)

shiny

METALS

magnetic (only iron, cobalt, nickel)

malleable (can be hammered into shape without cracking)

ductile (can be drawn into a wire)

high melting point

General properties of non-metals

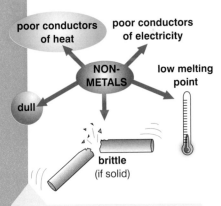

poor conductors of heat

poor conductors of electricity

NON-METALS

low melting point

dull

brittle (if solid)

ⓐ Look carefully at the picture. How can we tell that iron is a metal and sulfur is a non-metal?

ⓑ Why are metals used for electrical wires but non-metals are not?

In English lessons we have dictionaries that list words and spellings. The Periodic Table of elements is a bit like a science dictionary. It lists all the elements that there are. You can make all the millions of different materials found in the universe from just these elements.

Some elements have very long names. To help us, scientists have symbols for every element so that we do not have to write the whole word. For example, the metal aluminium has the symbol Al. (You will find out more in Fusion 2.)

Iron is a metal and sulfur is a non-metal

H Hydrogen																	He Helium
Li Lithium	Be Beryllium											B Boron	C Carbon	N Nitrogen	O Oxygen	F Fluorine	Ne Neon
Na Sodium	Mg Magnesium											Al Aluminium	Si Silicon	P Phosphorus	S Sulfur	Cl Chlorine	Ar Argon
K Potassium	Ca Calcium	Sc Scandium	Ti Titanium	V Vanadium	Cr Chromium	Mn Manganese	Fe Iron	Co Cobalt	Ni Nickel	Cu Copper	Zn Zinc	Ga Gallium	Ge Germanium	As Arsenic	Se Selenium	Br Bromine	Kr Krypton
Rb Rubidium	Sr Strontium	Y Yttrium	Zr Zirconium	Nb Niobium	Mo Molybdenum	Tc Technetium	Ru Ruthenium	Rh Rhodium	Pd Palladium	Ag Silver	Cd Cadmium	In Indium	Sn Tin	Sb Antimony	Te Tellurium	I Iodine	Xe Xenon
Cs Caesium	Ba Barium	La Lanthanum	Hf Hafnium	Ta Tantalum	W Tungsten	Re Rhenium	Os Osmium	Ir Iridium	Pt Platinum	Au Gold	Hg Mercury	Tl Thallium	Pb Lead	Bi Bismuth	Po Polonium	At Astatine	Rn Radon
Fr Francium	Ra Radium	Ac Actinium															

The Periodic Table is like a science dictionary

Classifying metals and non-metals

activity

You are going to investigate the properties of different materials to classify them as metals or non-metals. Think carefully about your results table to record your observations.

- Take a magnifying glass and look closely at the surface of the material.
- Put your sample into water. If it floats, it is less dense than water. If it sinks it will be more dense than water.
- Connect the material in a simple series circuit with a lamp and power supply. Turn on the power. Is the material a conductor of electricity?
- Tap the material on the side of the bench. How would you describe the sound?
- Place the material on a board and hit it with a hammer, but not too hard. What happens to the material? Eye protection needed.
- Note down your observations and decide which samples are metals and which are non-metals.

⚠ **Safety:** Crocodile clips can get hot. Keep power supply away from water.

Did You Know?

Mercury is the only metal that is liquid at room temperature. It is used in thermometers, as mercury expands by the same amount per 1°C temperature rise. If the glass holding the mercury breaks, then a special clean-up process with sulfur must be used, as the mercury is toxic.

science@work

Knowing the properties of materials is very important for architects and builders. If they put the wrong material into a building it could collapse!

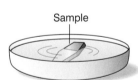
Sample

Sample

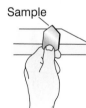
Sample

Sample

Summary Questions

❶ Copy and complete the sentences by matching the two columns:

a)	Most materials can be grouped	which means that they bend easily.
b	Carbon is a non-metal	as metals or non-metals.
c)	Metals are malleable	but it can conduct electricity.
d)	Mercury is	brittle, dull and insulators.
e)	Non-metals are	the only liquid metal at room temperature.

❷ Draw a table to compare the properties of metals and non-metals.

❸ Silver is the best electrical conductor, but copper is usually used in wires. Find out why silver is never used in wires.

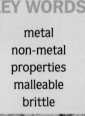

Burning

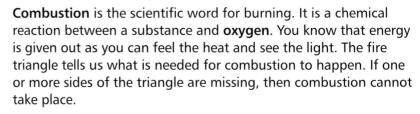

▶▶ How do we use a Bunsen burner safely?

▶▶ What is needed for burning?

▶▶ What happens when something burns?

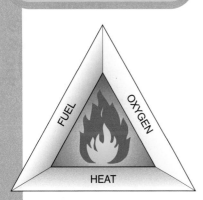

For combustion to happen you need fuel, oxygen and heat

Combustion is the scientific word for burning. It is a chemical reaction between a substance and **oxygen**. You know that energy is given out as you can feel the heat and see the light. The fire triangle tells us what is needed for combustion to happen. If one or more sides of the triangle are missing, then combustion cannot take place.

ⓐ What is the fuel used for a barbeque?

ⓑ When a log fire is burning, where does the oxygen come from?

ⓒ When birthday candles are burning, where did the heat come from to start off the reaction?

The fire triangle reminds us of what is needed for combustion to happen. By removing any side of the fire triangle, a fire will stop.

ⓓ What is removed from a chip-pan fire when it is covered with a fire blanket?

You will use a **Bunsen burner** in a science lab to heat things. Combustion happens between natural gas (methane) and oxygen from the air. You can change the heat from the Bunsen burner by twisting the collar so that the air hole is open or closed. The safety flame is the coolest flame and yellow. You should always leave the Bunsen burner on the safety flame if you are not using it. ▷

Combustion is part of a group of chemical reactions called oxidation reactions. This means that oxygen is being added to a chemical. Rusting and respiration are other examples of an oxidation reaction, but they are not called combustion reactions.

Fuels are chemicals that we burn or combust to release heat energy. Other chemicals can also combust, even if they are not fuels. Coloured chemicals are put into fireworks and they combust, but the heat energy is not used. This means that these chemicals are not fuels.

Another year older, another combustion reaction!

A Bunsen burner on the safety flame

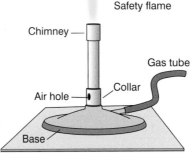

Chimney

Safety flame

Gas tube

Collar

Air hole

Base

Flame-proof mat

Combustion

Your teacher will demonstrate the combustion of iron, magnesium and carbon.

What do you observe? Predict the name of the product made.

⚠ What safety measures should your teacher take?

science @ work

You can now get special flooring that stops fires. As the special plastic gets hot, it releases a gas which stops the fire. The floor is expensive, but it may be cheaper than paying for water damage from sprinklers and the fire service.

When the fire-fighter hoses a fire, they are removing heat

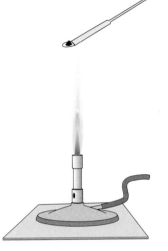

The chemical reaction of combustion can be shown in a word equation:

$$\text{chemical} + \text{oxygen} \rightarrow \text{chemical oxide}$$

Summary Questions

❶ Copy and complete with the words below:

combustion   oxygen   air   triangle   fuel

Combustion is an irreversible, chemical change between a substance and … .The fire … tells us the three things that are needed for combustion to happen: heat, … and oxygen. When a Bunsen burner is lit, … is taking place at the top of the chimney. The oxygen comes from the … , the gas is the fuel and the heat to start off the reaction is provided by a lighted splint.

❷ Copy and complete the following word equations. The general equation in 'Did you know?' will help you.

a)   sulfur + … → sulfur oxide

b)   aluminium + oxygen → …

c)   … + oxygen → sodium oxide

❸ Make a poster to explain how to use a Bunsen burner safely.

KEY WORDS

combustion
oxygen
Bunsen burner

Fuels and Oxygen

▶▶ What is a fuel?

▶▶ What are the products of combustion?

▶▶ What effect does oxygen have on a fire?

▶▶ How can I record the results of my investigation?

Fuel the fire

A **fuel** is a store of chemical energy. Fuels are burned to release their chemical energy into a useable form, mainly as heat. Fossil fuels such as coal, oil and natural gas, are made up mainly of **hydrocarbons**. When they react with oxygen they form carbon dioxide and water.

Hydrogen can be used as a fuel for cars

Great Debates

Cars use non-renewable fuels such as diesel and petrol. We get them from the fossil fuel, crude oil. These fuels could run out within your lifetime. There are alternatives such as using alcohol, hydrogen or biodiesel. What do you think the best fuel for a car is and why?

activity

Combustion of a hydrocarbon

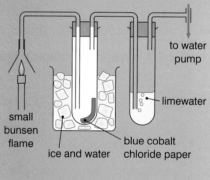

to water pump

limewater

small bunsen flame

ice and water

blue cobalt chloride paper

Your teacher will show you the burning of a hydrocarbon. The pump will pull through the products of combustion into the two different boiling tubes so that they can be tested. We use cobalt chloride paper to test for water. It turns from blue to pink if water is present.

Stretch Yourself

Why should you run the experiment without burning the fuel to prove what the products of combustion really are?

What would eventually happen to the limewater if you ran the experiment without the fuel burning?

a What is the hydrocarbon in this experiment?

b What happens to the limewater if carbon dioxide is made?

c Why is the first boiling tube in an ice bath?

Chemicals that burn easily are called **flammable**. They have a special hazard symbol. You may have seen this hazard symbol on chemical tankers, or even on solvents at home.

HIGHLY FLAMMABLE

activity

Investigating combustion

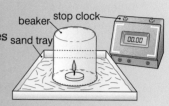

beaker — stop clock

sand tray

If a chip-pan is on fire, do not add water to it as this causes a flash of flame and the fire can spread. To stop a chip-pan fire, you turn off the heat to the pan. Then reduce the amount of oxygen by putting the lid on the pan, if the flames are not too big, or a damp tea towel over the fire.

You are going to investigate how the volume of air affects the burning of a candle. Make a **prediction** for this investigation. A prediction should be what you think is going to happen and why – using science to explain.

- Choose three different-sized glass containers.
- Fill each container with water.
- Using a measuring cylinder, measure the volume of water in each container.
- Light a candle in the centre of a sand tray.
- Put one of the empty containers upside-down over the candle and time how long it takes until the candle goes out.
- Will you need to repeat your test to make it more reliable? If so, how many times?
- Record your results in a table, showing any repeat readings taken.
- Do the same thing with the three other containers.
- What do you results show you? Was your prediction correct?
- Could your investigation be improved? Explain your answer.

> **Safety:** Remove things that catch light easily from the bench. Be careful not to burn yourself. Tie back hair and clothes.

---

① Copy the key words below and match them to their correct definitions:

a) fuel	a type of fuel; coal, oil and gas are examples
b) hydrocarbon	describing what we think will happen in an experiment and why
c) flammable	a chemical that we burn to release energy in a useful form, often as heat
d) prediction	a chemical that burns easily and has a hazard symbol

② Design a label for paint stripper – a flammable and harmful chemical.

Summary Questions

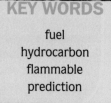

KEY WORDS

fuel
hydrocarbon
flammable
prediction

Making Oxygen

>> What do we observe when we make oxygen?

>> How do we test for oxygen?

All about oxygen

Oxygen makes up about 20% of dry air and is needed for life, as we know it, on Earth. Chemical reactions can make oxygen. For example, advertisements for some cleaning products boast that they contain oxygen bubbles to help the cleaning process.

Oxygen bubbles are made as the cleaner reacts

Space rockets have to carry their own oxygen to burn their fuel. This is because the further up in the atmosphere you go, the less oxygen there is. In space there is no oxygen that could be used for combustion. What type of chemical do you think rockets would carry as well as their fuel? What hazard symbols would they have on the tanks? What could be some safety concerns during take-off?

ⓐ What fraction of dry air is made up of oxygen?

ⓑ Why do you think that the amount of oxygen in the air is measured in dry air?

Did You Know?

You can buy hydrogen peroxide from the pharmacy to treat throat infections, as it kills microbes. In light, hydrogen peroxide breaks down to make water and oxygen. If you gargle with hydrogen peroxide, the oxygen bubbles push out pus from your throat and make it easier to spit up!

Oxygen gas can be collected by **displacement**. This is when a test tube is filled with water and held upside down in a water trough. A delivery tube takes the oxygen made in a chemical reaction into the test tube. As the oxygen is less dense than water, it rises to the top of the test tube and pushes (displaces) the water out. We can test for oxygen gas with a glowing splint. Oxygen will re-light the glowing splint.

Did You Know?

Manganese dioxide is a **catalyst**. This means it speeds up a chemical reaction but doesn't get used up itself. So, if you keep adding hydrogen peroxide more oxygen would be made, without having to add any more manganese dioxide.

Collecting oxygen under displacement

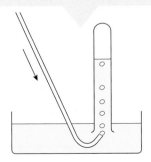

Testing for oxygen

Making oxygen

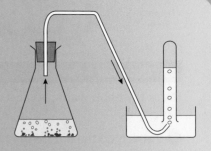

Hydrogen peroxide will break down into water and oxygen in normal sunlight. This is why the chemical is always stored in a brown or black bottle away from the light. This chemical reaction is very slow, but can be speeded up by adding manganese dioxide.

- Wear eye protection.
- Using a measuring cylinder put 25 cm³ of the hydrogen peroxide into a conical flask.
- Add half a spatula of manganese dioxide (HARMFUL).
- Quickly put the bung and delivery tube in place.
- Collect the gas under displacement.
- When a test tube has been filled, put the bung on under the water and put it into a test tube rack.
- Test the gas to prove that it is oxygen.

Safety: Manganese dioxide is harmful. Hydrogen peroxide can cause burns. Wear eye protection and wash your hands at the end of the practical.

If a chemical contains a lot of oxygen, which it can give up during a chemical reaction, it is called an **oxidiser**. An oxidiser can increase the speed of combustion and turn it into an explosion. Oxidisers also allow combustion to happen even if no oxygen gas is present. Oxidisers have a special hazard symbol.

OXIDISING
These substances provide oxygen, which allows other materials to burn more fiercely

✚ Help Yourself

Fold a piece of A4 coloured paper into three. Open out the paper and write hydrogen at the top of the first column, carbon dioxide at the top of the next column and oxygen at the top of the last column. Draw a labelled diagram and put a sentence in each column to remind you how to test for that gas. Look back at pages 66–69 to help you.

C Why should oxidisers and fuels never be stored in the same place?

Summary Questions

1 Copy and complete the sentences below:
   About one fifth of dry air is made …
   Oxygen gas can be tested …
   Some chemicals contain a lot of oxygen …
   Hydrogen peroxide is an example …

2 When hydrogen peroxide is exposed to light it breaks down into water and oxygen. Write a word equation for this reaction.

3 Find out some examples of oxidisers and what they are used for.

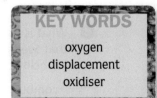

KEY WORDS

oxygen
displacement
oxidiser

know your stuff

▼ Question 1 (level 4)

Hazard symbols warn people if chemicals are dangerous.

(a) Match up the hazard symbols with their correct meanings: [3]

A   B   C

| toxic | flammable | irritant |

(b) Rachel was in the school science room and saw a bottle of dilute hydrochloric acid.

Which hazard symbol would be on the bottle? Choose from the list below:

     toxic   flammable   irritant   [1]

(c) Give one safety rule for using acid. [1]

▼ Question 2 (level 5)

Syed used universal indicator to test some chemicals that he found at home.

(a) Copy and complete Syed's results table:

Chemical	Colour	pH
Water	green	
Soap		10
Stomach acid		1
Lemon juice	yellow	
Sodium hydroxide	purple	

[5]

(b) In Syed's experiment, which substances were acids? [2]

(c) When Syed mixed the sodium hydroxide and stomach acid there was a chemical change. How did he know a reaction had happened? [1]

(d) What is the name of the chemical reaction that Syed had observed? Choose from the list below:

     combustion   oxidation
     neutralisation   rusting   [1]

▼ Question 3 (level 6)

Sandra put a piece of magnesium into a test tube of nitric acid.

(a) What type of change takes place? [1]

(b) What observation would tell Sandra a new substance has been made? [1]

(c) Sandra trapped the gas and tested it. She found that a popping noise was made with a lighted splint. What is the name of the gas that was made? [1]

(d) Sandra repeated her experiment with copper, zinc and calcium. She found that there was no fizzing with copper. Zinc made a few bubbles and calcium made lots of bubbles. Calcium is the most reactive of the four metals and copper is least reactive. Where would zinc and magnesium come in this order? [2]

▼ Question 4 (level 6)

We use a Bunsen burner in a science lab to heat chemicals.

(a) What is the name of the chemical reaction happening in the Bunsen burner? [1]

(b) The fuel in a Bunsen burner is methane. Copy and complete the word equation for the chemical reaction of the blue flame.

     … + oxygen → carbon dioxide + …   [1]

(c) One of the products of the blue flame is carbon dioxide. How can you test for this gas? [2]

(d) Why is the Bunsen burner flame hottest when the air hole is open? [1]

How Science Works

▼ Question 1 (level 4)

Adam investigated how the amount of air in a jar affects the time a candle burns for.

(a) Which piece of apparatus would he use to measure the time? Choose from the list below:

measuring cylinder   thermometer
          stop-watch   [1]

(b) What is the independent variable in Adam's experiment? (This is the variable which Adam has chosen to change in each test.)[1]

(c) What is the dependent variable in Adam's experiment? (This is the variable that Adam used to judge the effect of varying the independent variable.)   [1]

(d) Give one control variable in Adam's experiment.   [1]

▼ Question 2 (level 5)

Sundeep investigated the reaction between metals and hydrochloric acid. He measured the volume of gas that three metals gave off in one minute. He found out that silver released no gas, zinc gave off $1 \, cm^3$ and $5 \, cm^3$ was made by magnesium.

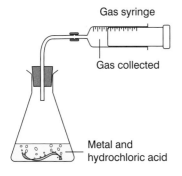

Gas syringe

Gas collected

Metal and
hydrochloric acid

(a) Copy and complete Sundeep's results table:

Metal	Volume of gas
Silver	0
	1
Magnesium	

[2]

(b) How could Sundeep best display his results? Choose from the list below:

bar chart   line graph   pie chart   [1]

(c) How could Sundeep make his work more reliable?   [2]

▼ Question 3 (level 6)

Emma investigated the volume of gas given off when calcium carbonate was added to hydrochloric acid. She recorded her results in the following table:

Time (s)	Volume of gas (cm^3)
0	0
10	15
20	28
30	17
40	40
50	42
60	43

(a) Copy and complete the graph below. Include the missing point and the labels on each axis.   [3]

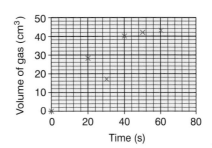

(b) Draw a line of best fit and circle the point that does not fit the pattern.   [2]

Particles in Action

What is matter?

- Everything is made of matter. Matter is made of tiny particles known as **atoms**.
- There are only about 100 types of naturally occurring atoms. These atoms join and mix in different ways to make all the materials that are in the universe!

- There are three states of matter:
  - solid
  - liquid
  - gas
- The particles in a substance arrange themselves in different ways to give each state of matter different properties.

a List three solids that are in your classroom.

b List three liquids that you could find in your kitchen.

c List three gases that you could find in the air.

Solids do not pour and do not take the shape of their container

Liquids can be poured and take the shape of their container

activity

Sorting out matter

You are going to investigate some chemicals. Using your observation skills (see page 54) you will group the chemical as a solid, liquid or a gas.

Your teacher will give you six samples labelled A to F for you to investigate. You may touch them and try to pour them. Record your results in a table, similar to the one below:

Sample	Observation	Solid, liquid or gas
A		
B		

Gases can be poured and fill up any container

The same substance can be a solid, liquid or a gas, just by changing its temperature. This is an example of a reversible, physical change. For example, below 0°C water is a solid (ice), between 0°C and 100°C it is a liquid, and above 100°C it is a gas (steam).

Water in its three states of matter

Solids

>> What is a solid?

>> How are the particles arranged in a solid?

You can't squash me!

Solid is one of the three states of matter. Solids have their own shape and volume. It does not matter what container you put a solid into, the shape stays the same. Solids cannot be squashed (compressed).

activity

Investigating solids

You are going to investigate the properties of solids by carrying out four experiments. You are going to write a **conclusion**, to say what you have found out in your experiment.

Compression

- Try to compress the wood in the syringe. What do you notice?

Bar and gauge

- Place the bar into the gauge, measure it and remove it.
- Set the Bunsen burner onto the blue heating flame.

  ⚠ **Safety:** Be careful not to burn yourself. Metals will stay hot for a long time.

- Heat the bar.
- Try to put the bar back into the gauge. What do you notice?

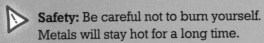

wood

gauge

bar

Ball and ring

- Try to put the ball through the ring.
- Heat the ball in the blue Bunsen flame.
- Try to put the ball through the ring.
- Now immerse the ball and ring in cold water. What do you notice?

  ⚠ **Safety:** Be careful not to burn yourself. Metals will stay hot for a long time.

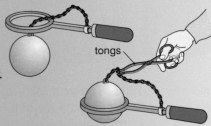

tongs

Mass

- Take three block of the same size.
- Compare the mass of each of these blocks. What do you notice?
- Try to explain the experiments above using your ideas about particles.
- Discuss your ideas with your partner, and then share your ideas with another group.

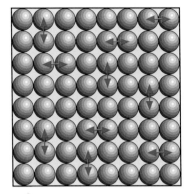

Solids have a fixed shape and volume

Particles in a solid have a regular arrangement and vibrate

The **particles** in a solid are lined up in a pattern. Each particle is touching its neighbours and they stay in this fixed arrangement. They cannot move around, but they do **vibrate** (bounce around on the spot). The hotter the solid is, the faster its particles vibrate. Eventually, the vibrations will be so strong that the particles begin to break free from their neighbours and the solid melts.

Did You Know?

Particles can only be still at absolute zero (–273°C). It has not been possible for scientists to cool any matter to this temperature yet.

a Why can't solids be compressed? Use the particle diagram to help you answer this question.

b Why can't solids be poured? Use the particle diagram to help you answer this question.

Summary Questions

1 Copy and complete using the words below (you can use each one more than once):

vibrate   expand   compressed

Solids have fixed shape and volume. They cannot be poured or … as the particles are in a regular pattern. The particles … in a solid. The hotter the solid is, the faster the vibrations. Solids … as they are heated. This is because the particles … faster and this makes the particles move slightly further apart. Each particle itself does not … .

2 What is the name of the process when a solid becomes a liquid?

3 What is the name of the process when a liquid becomes a solid?

4 Make a particle model of a solid. It could be a drama, sculpture or a diagram. Then look carefully at your model. What are its good points and what are its bad points?

KEY WORDS

solid
conclusion
particle
vibrate

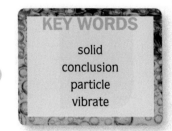

Liquids

C2.3

▸▸ What is a liquid?

▸▸ How are the particles arranged in a liquid?

Wet wet wet

You have looked at the solid state of matter in some detail. **Liquid** is another state of matter. Liquids have a fixed volume, but they can change their shape. The same liquid will have a different shape in different containers. Liquids can only be squashed (compressed) a very small amount – so small that you would not notice.

activity

Investigating liquids

You are going to investigate the properties of liquids by completing three different experiments. You are going to write a conclusion.

Compression

- Try to compress the water in the syringe. What do you notice?

  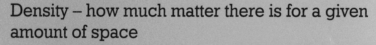
  water

  ⚠ **Safety:** Always point a syringe away from faces.

Home-made thermometer

- Put the conical flask into a hot water bath.
- Look carefully at the level of liquid in the glass tube. What do you notice?

  ⚠ **Safety:** Wear eye protection.

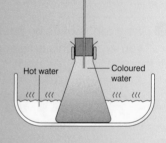
Thin glass tube

Hot water
Coloured water

Density – how much matter there is for a given amount of space

- Pour some oil into a test tube.
- Add some water and place a bung in the top.
- Shake the test tube, then leave it for a few minutes.
- What do you notice when you look at it again?
- Try to explain the experiments above using your ideas about particles.
- Discuss your ideas with your partner, and then share your ideas with another group.

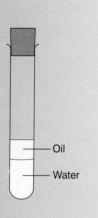
Oil

Water

a Why do you think kettles have a space between the top of the water in the kettle and the lid?

The particles in a liquid are very close together but they can move past each other. This means that they have a **random** arrangement. At least half of the particles are touching at any time and they take up slightly more space than in a solid. The hotter the liquid is, the faster the particles move. As the temperature rises more and more particles gain enough energy to escape from the liquid. Eventually, the liquid boils and becomes a gas.

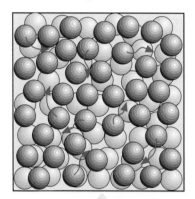

Particles in a liquid have no pattern and can easily move past each other

Did You Know?

Glass is actually a liquid! If you look carefully at windows in old buildings you will see that the glass has flowed downwards because of gravity. This means that the bottom of the glass is thicker than the top. The particles in a piece of glass can move past each other. Because they are in a supercooled liquid, this movement happens very slowly.

b Why can liquids be compressed but you are unlikely to see it?

Summary Questions

1 Copy and complete the sentences by matching the two columns:

a)	Liquid particles have an	and the gap between the particles gets bigger, but each particle doesn't change shape.
b)	When a liquid is heated, the particles move faster	poured because they do not have a fixed shape.
c)	Liquids can be	when they are heated.
d)	Liquids expand	irregular arrangement and can easily move past each other.

2 Make a table to compare a solid and a liquid using the particle model. List all the similarities and all the differences.

3 Write a short story to explain what it would be like to be a particle in a solid which melts to become a liquid.

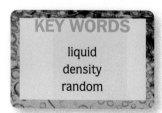

KEY WORDS

liquid
density
random

Gases

>> What is a gas?

>> How are the particles arranged in a gas?

I want to break free

Gas is a state of matter. Particles in the gas state have more movement energy than the same amount of a substance in the liquid or solid state. Gases spread out into any container that they are put in. This means that they have no fixed shape or volume. Gases can be squashed (compressed) by a large amount.

activity

Investigating gases

You are going to investigate the properties of gases by completing three different experiments.

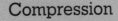
air

Compression

- Try to compress the air in the syringe. What do you notice?

 Safety: Always point a syringe away from faces.

Appearance

- Look carefully at the different samples of gases in gas jars.

- What do you notice?

Mass

- Put a beaker on to an electronic balance and press zero.

- Pour some carbon dioxide gas into the beaker.

- What do you notice?

- Try to explain the experiments above using your ideas about particles.

- Discuss your ideas with your partner, and then share your ideas with another group.

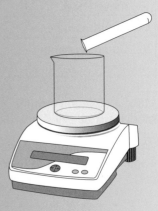

The particles in a gas take up all the space of a container. They move very fast and in any direction. This means the particles have a random arrangement. The hotter the gas is, the faster the particles move, and the **pressure** of the gas in a sealed container is higher. The pressure of a gas is caused by the particles hitting the sides of the container. The more hits there are in a set time, the higher the pressure.

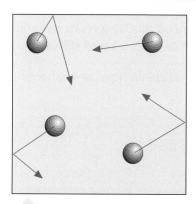

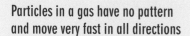

Particles in a gas have no pattern and move very fast in all directions

a Using the particle model of a gas to explain why a gas can be compressed so easily.

b Explain how you know that the particles in a gas have more energy than particles of the same substance as a liquid or solid. Use the particle model of a gas to help you.

Did You Know?

Gases can be compressed so much that they actually become a liquid. These gases are describes as being 'under high pressure'. Examples of gases under very high pressure include camping gas.

Summary Questions

1 Copy the key words below and match them to their correct definitions:

a) gas	a variable measured in kilograms (kg)
b) mass	the force on a set area caused by gas particles hitting the sides of the container
c) pressure	a state of matter, where the particles have the most energy

2 Make a list of the properties of gases and explain them using the particle model.

3 One way to increase gas pressure is to heat the gas. How else could you increase gas pressure? Explain your answer.

KEY WORDS

gas
pressure

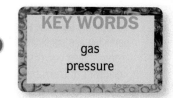

Gases in Action

▸▸ How do gases spread out?

▸▸ How strong is gas pressure?

Catch me if you can

Gas particles are moving in all directions all the time. This means that when a gas is released into a new container it quickly spreads out and fills the new container. The spreading out of a gas is called **diffusion.**

a Draw a labelled particle diagram to explain how air freshener spreads through a room.

Did You Know?

Diffusion can also happen in liquids, but it is slower. This is because liquid particles can also move and they can diffuse.

This green dye is diffusing through water

activity

Diffusion in gases

Your teacher will put a little bit of bromine at the bottom of a gas jar and then place a second gas jar on top of it. Bromine is a dense brown liquid, but at room temperature it evaporates to make an orange gas.

● What do you predict will happen? Don't forget to explain your prediction using the particle model.

All around us is air. Air is a mixture of gases, mainly nitrogen and about one-fifth oxygen. The particles of these gases are constantly hitting our bodies, and they cause **air pressure**. As the particles are hitting us all the time, we do not realise the force on our bodies.

b How can you find out what the air pressure is at the moment?

If something is difficult to imagine, then a **model** can help us understand and explain observations. Models show simplified ideas, so they always have drawbacks because they cannot explain every situation.

Next time you...

... spray an air freshener, try to imagine how the smelly gas particles are spreading to fill the room. Could you draw a particle diagram to show this happening?

I thought I felt under pressure!

Air pressure

You are going to investigate the force of air pressure on an empty can.

- Fill a trough with cold water.
- Take an empty drinks can and put a small amount of water at the bottom.
- Fix a clamp around the can.
- Heat the can on a blue Bunsen burner flame, until you see steam coming out of the top of the can.
- Turn the can upside down, so that the drinking hole is in the water trough. What do you observe?

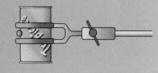

Water

- Explain your observations using the particle model.

 Safety: Do not burn yourself on the can, steam or the Bunsen burner.

Did You Know?

Hyperbaric chambers increase or decrease the number of gas particles in a small air tight chamber. This increases or decreases the pressure on the body and changes how easy it is to breathe and the amounts of dissolved gases in our bodies. These chambers can be used in medical treatments.

Stretch Yourself

Half fill a beaker with custard powder. Add just enough water to make a paste. Quickly hit the surface of the custard – what state of matter is this? Now try to pour the custard – what state of matter is this?

Summary Questions

1. Copy and complete the sentences below:

    Gases spread out ...

    Air particles move in all directions and ...

    A model is ...

2. Look at the diagram of a neutralisation reaction between hydrogen chloride gas (acid) and ammonia gas (alkali). When they meet they make a white solid.

Hydrogen chloride

Ammonia

   Explain how the gases meet each other. Suggest why the solid is not formed in the centre of the tube.

3. Make a pictorial flow chart to explain how the can collapsed in the above activity because of air pressure.

KEY WORDS

diffusion
air pressure
model

Changing State

▶▶ What happens when we heat a solid or a liquid?

▶▶ What happens when we cool a liquid or a gas?

Physical changes

Most substances can be made into a solid, liquid or gas, you just need to cool or heat them. Changing state is reversible and an example of a physical change. (You will learn more about this in Year 8.)

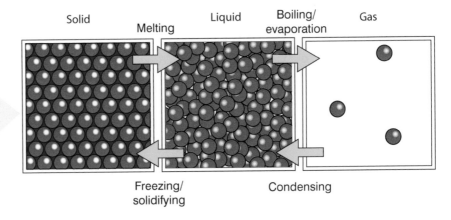

Changing state is a physical change

> **a** Why is changing state a physical change?

> **b** What is the name of the physical change when a gas turns into a liquid?

There are two ways to change from a liquid to a gas:

* **evaporation**
* **boiling**

In evaporation, the liquid particles bounce off each other and sometimes one particle gets hit by lots of other particles and it gains enough energy to escape from the surface of the liquid as a gas. The remaining liquid then gets colder. Evaporation happens below a substance's boiling point.

In boiling, energy is given to all the particles at the same time by an outside source, like a cooker or Bunsen burner. Boiling happens at the substance's boiling point, and the temperature remains the same until all of the liquid has turned into a gas.

> **c** Give an example of evaporation that happens in everyday life.

When we cool a gas or a liquid, movement energy is lost from the particles. If we cool a gas, eventually the particles move close together and slower. Eventually the gas **condenses** into a liquid.

Did You Know?

Another scientific word for 'melt' is 'fuse'. In a plug there are fuses – named after the physical change of melting. These are thin wires in glass tubes. If the current gets too high, the metal wire heats up and melts (or fuses) and stops the electricity. This makes the machine safe.

✚ Help Yourself

Look back at C2.2, C2.3 and C2.4. Make sure that you can draw the particle diagrams for the different states of matter. You might even want to try to draw a particle diagram to explain the changes of state.

If we cool a liquid, movement energy is lost by the particles and they move less and less. Eventually the particles only vibrate in a fixed position. The liquid solidifies or **freezes** to become a solid.

Iodine can change state from a solid to a gas without being a liquid

Stretch Yourself

It is possible to go straight from being a solid to a gas without ever being a liquid. Iodine crystals are a grey solid at room temperature. If you put solid iodine in a sealed test tube and heat it, you will see purple gas, but never a liquid. What is this physical change called? What is the reverse of this physical change called (going from a gas to a solid without being a liquid)?

Next time you...

...are cooking, how many changes of state can you notice? Maybe the butter melting, or water boiling.

Summary Questions

1 Copy the key words below and match them to their correct definitions:

a) melting	a physical change where a solid becomes a liquid
b) evaporation	a physical change where a liquid becomes a solid
c) boiling	a physical change where a gas becomes a liquid
d) condense	a physical change where a liquid below its boiling point becomes a gas
e) freeze	a fast physical change where a liquid becomes a gas

2

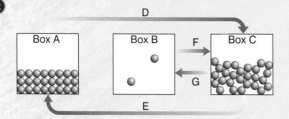

Look carefully at the diagram above and use the letters to answer the questions below:

a) Which letter shows the particles in a gas?

b) Which letter shows the particles that cannot be compressed?

c) Which letter shows evaporation?

d) Which letter shows boiling?

e) Which letter shows freezing?

3 Draw a cartoon strip to explain what it would feel like to be a steam particle that is cooled to form part of an ice cube.

Did You Know?

Evaporation is one way that our bodies regulate our temperature. Our sweat glands leak liquid onto our skin, the heat from our blood helps to make the sweat evaporate, taking some of the heat energy away and cooling our bodies.

Thank goodness for evaporation!

KEY WORDS

melting
evaporation
boiling
condense
freeze

Mixtures

- ▶▶ What is a mixture?
- ▶▶ How are the particles arranged in a mixture?

Pure and simple

When scientists describe water as **pure**, they mean that it only contains water particles. Some tap water contains dissolved minerals from the rocks that the water has flowed over.

Water companies add dissolved fluoride to help make our teeth strong and dissolved chlorine to kill microbes. Gases from the air will also dissolve into the water. This means that tap water is not pure, it is a **mixture**.

A mixture is made up of two or more substances that are not chemically joined together.

ⓐ A bottle of mineral water has 'pure water' on its label. Would a scientist describe this as pure water? Explain your answer.

ⓑ When a mixture is made, is this a chemical or physical change? Explain your answer.

We can use the particle model to show a mixture. Brine is an example of a special type of mixture called a **solution**. Brine is a mixture of salt (sodium chloride) and water. The salt particles fit in the gaps between the water particles.

The scientific word for the substance that will dissolve is **solute** and the liquid that does the dissolving is the **solvent**.

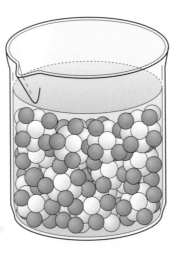

Brine is a mixture

Investigating mixtures

You are going to investigate the mass and volume of brine.

- Measure 100 cm³ of distilled water into a measuring cylinder.
- Measure the mass of the water on an electronic balance.
- On an electronic balance, measure 1.0 g of salt into a beaker.
- Predict what the mass and volume of the mixture of salt and water will be. Explain your prediction using science.
- Add the water to the beaker and stir until you can no longer see any salt.
- Take the mass and volume of the mixture.
- Record your results in a table.
- Look at the evidence and decide if your prediction is correct or not.

activity

When no more solute will dissolve, the solution is described as 'saturated'. Substances that will not dissolve in a particular solvent are said to be 'insoluble' in that solvent.

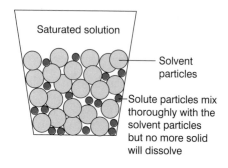

Saturated solution

Solvent particles

Solute particles mix thoroughly with the solvent particles but no more solid will dissolve

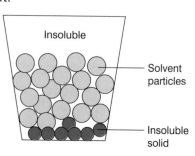

Insoluble

Solvent particles

Insoluble solid

Summary Questions

1. Copy and complete using the words below:

   mixture   solvent   solution   gaps

   A … is more than one substance not chemically joined. Solutions are a group of mixtures made up of solute particles that mix completely with the … particles. When no more solute can dissolve, we say that the … is saturated.

2. Find out some other examples of groups of mixtures.

KEY WORDS

pure
mixture
solution
solute
solvent

Separating Mixtures: Sieving and Filtering

> ▶▶ How can we separate different-sized pieces of solid?
>
> ▶▶ How can we separate solids from liquids?

A mixture is made up of two or more substances that are not chemically joined. This means that we can separate mixtures into their parts using physical means. The method for **separating** a mixture depends on what the mixture is made from.

You might have used a **sieve** at primary school to separate sand and stones. This is because the grains of sand are small enough to fall through the sieve, but the stones are too big. Sieving can be used to separate different-sized pieces of solids.

> Sieving can be used to separate different-sized pieces of solids

Did You Know?

Filter paper is graded. The number on the box tells you how small the holes are in the paper. The smaller the holes, the more expensive the paper is.

ⓐ How could you separate a mixture of sand, small beads and large beads?

ⓑ How could you separate salt from sugar cubes? Explain your answer.

Filtering can be used to separate an insoluble solid from a liquid. The **filter** paper is like a very fine sieve, where the holes are too small to see unless you have a powerful microscope.

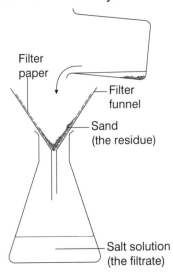

Filter paper

Filter funnel

Sand (the residue)

Salt solution (the filtrate)

> Filtering can be used to separate an insoluble solid from a liquid

C How could you separate sand and water?

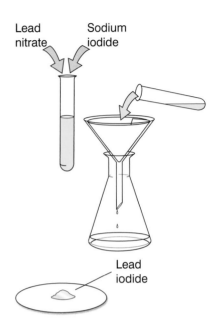

Lead nitrate   Sodium iodide

Lead iodide

Separating mixtures

You are going to separate an insoluble solid from a solution.

- Using a measuring cylinder, measure $5\,cm^3$ of the lead nitrate solution into a test tube.
- Add $10\,cm^3$ of the sodium iodide solution.
- What do you observe?
- Rinse the filter paper with water.
- Filter the mixture. The liquid (**filtrate**) is sodium nitrate and can be put down the sink.
- The solid on the paper is lead iodide, an insoluble solid.
- Why must the lead iodide be filtered rather than sieved from the mixture?

⚠ **Safety:** Wear eye protection and wash hands after this practical. Lead iodide is toxic.

❶ Copy the key words below and match them to their correct definitions:

a) separating	a method for separating different-sized pieces of solid
b) sieve	a method for separating an insoluble solid from a liquid
c) filter	the liquid that is collected after filtering
d) filtrate	sorting a mixture into the parts that it is made from

❷ Explain how you would separate sand grains from salt grains of the same size.

❸ Research to find out how many different grades of filter paper there are and some of their uses.

Summary Questions

KEY WORDS

separating
sieve
filter
filtrate

Separating Mixtures: Chromatography

▸▸ How can we separate different inks and dyes?

▸▸ How can we use chromatography?

The colours used in pens and foods are often mixtures of more than one ink or **dye**. These colours can be made from natural things like plants and animals, or made by scientists in a lab.

Chromatography is a way to separate the different colours in a dye. A **chromatogram**, meaning 'coloured picture', is made. A chromatogram shows the separate colours that make the mixture.

ⓐ Predict what a chromatogram of a pure dye would look like.

activity

Making a chromatogram

You are going to find out how many colours make up black ink.

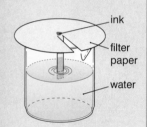

- Take a round piece of filter paper, put a drop of black ink in the centre and allow it to dry.

- Cut a wick (a rectangle) into the centre of the paper – be careful not to cut away the wick.

- Half fill a beaker with water.

- Bend the wick so that it is in the water.

- Observe as the colours separate.

- When the water line is past the coloured circles, your chromatogram is finished.

- Remove the chromatogram and allow it to dry in a warm place.

- How many colours made up your black ink?

Most dyes are actually solutions of different coloured chemicals. The colours are the solute particles. They are attracted by the solvent particles and the chromatography paper. Each solute is attracted by a different amount. When the colour is separated by chromatography, the solutes that are held strongest by the solvent travel the furthest, those that are held the strongest by the paper travel the least.

ⓑ Which solvent do you think we use mostly in schools for chromatography?

Don't worry! With chromatography we'll find the person who did this!

We can use chromatography to compare an unknown dye with known chemicals to find out its name.

Using chromatography

You are going to find out which food colouring you have.

- Take a rectangular piece of chromatography paper and draw a pencil line about 1 cm from the bottom edge.
- Put three crosses on the pencil line, leaving about 2 cm between each cross. Label the first cross as sample and the next two as A and B.
- Drop a sample of the unknown food colouring onto the first cross.
- Drop a sample of food colourings A and B onto each of their crosses.
- Roll the paper and clip it into a roll.
- Put about ½ cm of water in the bottom of a beaker.
- Lower the paper into the water, so that the pencil line is closest to the bottom, making sure the water is not above the pencil line.
- Observe until your chromatogram has finished developing and allow it to dry.
- Which food colouring was the unknown sample? How do you know?

paper clip
solvent front
water (the solvent)
A   B

c Why do you think pencil is used rather than pen to make a chromatogram?

Link up to ...

PE
Sports stars are often tested to make sure that they are not taking drugs and cheating. They give a urine (wee) sample and chromatography is used to find out what chemicals are in the urine. Drugs can be identified and the sports person may be banned from competing.

Summary Questions

1 Finish the following sentences:
Chromatography can be used …
A chromatogram is …
Many inks and dyes …

2 Millie made a chromatogram of orange squash and compared the colouring used with banned colours.

solvent front

orange squash   A   B   C   D   E
banned dyes   safe dyes

   a) How many dyes are used in the orange squash?

   b) Which dyes have been used in the orange squash?

   c) Is the orange squash safe to drink? Explain your answer.

3 Chromatography can be used to see the DNA of humans. Find out how electricity is used to separate DNA.

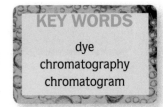

KEY WORDS
dye
chromatography
chromatogram

Separating Mixtures: Distillation and Evaporation

▶▶ How can we separate a solid dissolved in a liquid?

▶▶ How can we separate a mixture of liquids?

Sometimes you might want to get a solute back from a solution. A solute can be separated from a solution using **evaporation**, but the solvent is lost to the air. (To remind yourself about evaporation see page 90).

activity

Evaporation

You are going to get salt from brine using evaporation.

- Half fill an evaporating basin with brine.
- Leave on a window ledge overnight.
- What do you predict will happen? Don't forget to explain your prediction using scientific ideas.
- Observe your evaporating basin the next day.
- Was your prediction correct?

Stretch Yourself

Air can be separated using fractional distillation. Air is cooled so that it condenses into a liquid. The liquid mixture is then slowly allowed to warm up so that when the boiling point of nitrogen in the air is reached, it turns into a gas and is condensed. The next gas to boil is oxygen; this is then condensed. Other gases like argon follow. Each gas that is collected is known as a fraction, as it makes up part of the mixture. Fractional distillation of air gives us important products that are used, e.g. oxygen in hospitals. Find out some other uses for fractional distillation.

ⓐ How could you speed up evaporation?

If we want to collect both the solvent and solute, then we can use **distillation**. Distillation can also be used to separate more than one liquid from each other. If the liquids have similar boiling points we use a technique called 'fractional' distillation.

Crude oil is made from tiny dead animals and plants that lived in the sea millions of years ago. Instead of rotting away, they made a rich mixture of chemicals called 'crude oil'. The mixture is not very useful, but when it is separated, each fraction (part of the mixture) has many uses, e.g. as fuels. Crude oil is separated using fractional distillation.

activity

Distillation

You are going to separate pure water from food colourings.

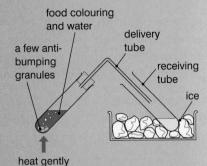

- Measure out 20 cm³ of coloured water into a boiling tube.
- Set up the apparatus as shown in the diagram.
- Using a Bunsen burner, heat the water gently and observe.
- What is left in the boiling tube?
- What is the **distillate** (the liquid collected in the receiving tube)?

b When could distillation be used in real life?

HISTORY

Many hundreds of years ago, Arabian scientists were the first people to use distillation. These early scientists were called 'alchemists' and their mission was to make gold from other simple substances. The alchemists never managed to make gold, but they did make a lot of useful discoveries.

Early distillation apparatus

Summary Questions

1 Copy and complete the following sentences by matching the two columns:

a)	Solvents can be separated and collected from	both the solvent and solute could be collected.
b)	In evaporation	a solution using distillation.
c)	In distillation	a type of mixture made of solute and solvent particles.
d)	Solutions are	only the solute is collected.

2 Look at the distillation apparatus.

a) What is the solute?

b) What is the solvent?

c) What is the distillate?

d) What is the job of the condenser?

3 Make a flow chart to explain the stages of distillation.

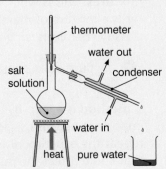

KEY WORDS

evaporation
distillation
distillate

Grouping Chemicals

» How can we decide if something is a solid, liquid or a gas?

» How can we decide if something is pure or a mixture?

» Can I devise a method to solve a problem?

Did You Know?

A duck-billed platypus lays eggs and has a duck's bill – so is it a bird? No, it is actually a mammal! This is an example of an animal that does not fit into a category easily.

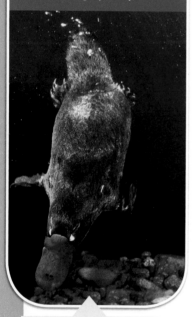

A duck-billed platypus

Where does it belong?

Scientists often want to group things, or classify them. This makes it easier to study them and, when new things are found, we can make predictions. However classifying is not always easy and sometimes things do not fit into any group or category.

One of the main **classifications** for chemicals is their state. However, some substances do not fit neatly into one state or another.

activity

Grouping chemicals by their state

You are going to be given three different mixtures. You need to decide if they are solid, liquid or gas, and why.

Copy and complete the table below for your results and conclusions:

What classification are these mixtures?

Mixture	Observations	Classification	Explanation
Tomato ketchup			
Hair gel			
Emulsion paint			

ⓐ What are the three states of matter?

Pure substances contain only one type of particle. This means that pure oxygen only has oxygen particles in it. Often chemicals are **mixtures**. This is when more than one type of substance is present. Air is a mixture of mainly oxygen and nitrogen. Another way to classify chemicals is by whether they are pure or a mixture.

ⓑ Is 'pure milk' scientifically pure? Explain your answer.

A **method** is a step-by-step guide for an experiment. Methods are important because scientists like to compare results. Others need to repeat the experiment, so that they do not do anything differently that could affect the results. Methods may use a diagram rather than trying to explain how to put the equipment together.

activity

Grouping chemicals as pure or impure

You are going to be given three samples. You must decide if they are pure or mixtures. The only way to do this is to try to separate the samples (see pages 94–99). If you cannot separate the chemical, then it is pure.

● Write a method for your practical. You may want to use diagrams to help explain some parts of the experiment.

● Copy and complete the table below for your results and conclusions:

Sample	Observations	Classification	Explanation
Ink			
Water			
White powder			

❶ Copy the key words below and match them to their correct definitions:

a) classification	only one substance
b) mixture	grouping things based on observations
c) pure	more than one substance, not chemically joined
d) method	a step-by-step guide to an experiment

❷ Rock salt is mined and purified to get pure salt from the sand and other insoluble impurities in the rock.

Write a method (step-by-step) guide to how you would separate salt from rock salt.

A rock-salt mine

Summary Questions

KEY WORDS

classification
pure
mixtures
method

know your stuff

▼ Question 1 (level 4)

Materials have different uses because of their different properties. Some properties of solids, liquids and gases are listed in the table. Copy the table below. Put in five ticks to show if the property is for a solid, liquid or gas.

Property	Solid	Liquid	Gas
Can flow			
Cannot be squashed			
Takes the shape of its container			

[5]

▼ Question 2 (level 5)

A water-soluble, black ink is an example of a mixture.

a What is a mixture? [2]

b What is the solvent in this mixture? [1]

c What process could you use to get pure water from the black ink? [1]

d A chromatogram was made of the black ink.

Explain how you know the chromatogram is finished. [1]

e How many dyes make up the black ink? [1]

▼ Question 3 (level 6)

A total of 200 cm³ of water was put into a beaker and left on a sunny windowsill.

At start   Two weeks later

a What process is happening to the water over the two weeks? [1]

b Which diagram below shows the water particles at the start of the experiment? [1]

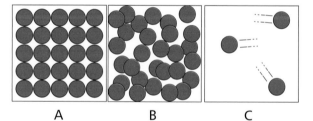

A   B   C

c How do the particles move differently in a gas compared to a liquid? [1]

d How do the particles in a gas move in a similar way to those in a liquid? [1]

How Science Works

▼ Question 1 (level 4)

Saffron and Baljit wanted to compare how quickly different-sized pieces of sugar would dissolve in water.

(a) What is the dependent variable? (This is the variable that Saffron and Baljit used to judge the effect of varying sizes of the pieces of sugar.)

Choose from the list below:

time   particle size   temperature

[1]

(b) What is the symbol for the unit of the dependent variable? Choose from the list below:

s   cm   °C   [1]

(c) Saffron and Baljit investigated the solubility of icing sugar, caster sugar and granulated sugar.

Draw a table for them to put their results into. [3]

(d) Below is a picture of Saffron's stop watch. This is the reading that she saw for the time it took for the granulated sugar to look like it had disappeared.

How many seconds did it take for the granulated sugar to dissolve? Give your answer to the nearest second. [1]

▼ Question 2 (level 5)

An anonymous letter was sent to Bob, from someone in his class. The letter was written in blue ink and only three other people in the class use blue ink.

(a) What technique could Bob use to separate the dyes in the inks? [1]

(b) Bob's method is written in the wrong order below. Copy and complete by putting the letters in order from 1 to 7. The first one has been done for you.

1 = G

A   Drop a sample of each ink onto each of the crosses.

B   Observe until the water level is above all the inks; then allow to dry.

C   Roll the paper and clip it.

D   Take an rectangular piece of chromatography paper and draw a pencil line about 1 cm from the bottom edge.

E   Lower the paper into the water, so that the pencil line is closest to the bottom.

F   Put four crosses on the pencil line.

G   Put about ½ cm of water in the bottom of a beaker. [6]

▼ Question 3 (level 6)

Lucy decided to investigate how the mass changes when copper sulfate dissolves in water. Below is a label on the bottle of the chemical:

copper sulfate

(a) Which hazard symbol is on the bottle? [1]

(b) Give one safety rule that Lucy should follow when using this chemical. [1]

(c) Lucy measured 100 g of water at 40°C into a beaker. She then measured 1 g of copper sulfate. Predict the mass of the copper sulfate solution. Explain your answer. [2]

(d) How could Lucy improve the precision of her mass readings? [1]

Electricity and Magnetism

Strike a light!

- Why do we find torches so useful? We can carry them around because they run on batteries. They are also ready for us to use at any time, in any place. There are hazardous chemicals inside the batteries, but they are carefully sealed in. This means you don't need to worry about them leaking in your pocket or bag. The bulb will last for many hours. Modern torches are often fitted with LEDs instead of bulbs. These make more efficient use of the electricity and can last for 10 000 hours.

- The advert shows how Ever-Ready torches were advertised a century ago. People had to be reassured that there were 'no wires, no liquids, no danger'. We take that for granted today.

This advert first appeared in 1904

- Electrical and electronic engineers produce new and exciting inventions every year. You don't have to understand how they work to enjoy using them.

- In this topic, you will build on what you already know about electric circuits. You will understand more about what goes on whenever you flick a switch and set electricity to work.

a Look at the advert for the Ever-Ready torch. It tells us things that we would take for granted today. Choose a modern electrical device (e.g. a MP3 player, CCTV camera, microwave oven) and make up an advert for it. Point out the features which would surprise someone living a century ago.

Switching on

- Try out some electric circuits.

- You should be able to remember how to draw circuit diagrams. Draw diagrams to represent some of the circuits you have tried out above.

- Make a list of all the different things we can use electricity to do.

b We can use electricity for many different purposes – for lighting, making things move, and so on. Think how you used electricity in the experiments above. Think about the ways you have used electricity so far today. Suggest a way in which it can make a big difference to the quality of people's lives.

c When a new invention comes along, such as mains electricity, people think of lots of uses for it. The 'light bath' shown on the right is an example. When the door was closed and the lights were switched on, the customer became very hot and sweaty. In Victorian times, this was advertised as a cure for asthma and other disorders. Plan a test to find out whether a light bath really would have benefits for people's health.

An unusual use for electricity – a 'light bath'

Complete Circuits

» How do we use conductors and insulators to make circuits?

» How do scientists represent an electric circuit?

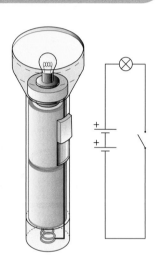

Lighting-up time

The photograph shows a light bulb – you can see the filament inside. When electricity passes through the bulb, the filament gets hot and starts to glow brightly. (Bulbs like this are becoming out-of-date, because they use too much electricity.)

A torch has a battery, a switch and a bulb. When you switch on the torch, electricity makes the bulb light up. So how does this work?

The diagram shows the insides of the torch. You can see that the battery has a plus (+) and a minus (-) sign, so that it is put in the right way round. You can also see the switch and the bulb, but it isn't clear how the circuit works.

The **circuit diagram** is much clearer. Each **component** in the circuit has its own special symbol. The lines in between show how the components are connected together.

The cat can't understand how this bulb lights up. Can you?

> **a** Draw the circuit symbol for each of the following components: battery, bulb, switch.
>
> Write the name of the component next to each symbol.

A special language

We use circuit diagrams to show how electrical components are connected together. The circuit diagram for the torch shows that, when the torch's switch is closed, there is a complete circuit.

- Put your finger on the positive (+) end of the battery.
- Use your finger to trace round the circuit until you get to the negative (–) end of the battery.

If there are no gaps, you know that the circuit is complete.

> **b** The torch circuit includes a battery, switch and bulb. Suggest another use for this circuit.

activity

Connecting up

- Try connecting some different electrical components in a circuit with a battery. Use lamps, switches, a buzzer and a motor.
- To connect up a circuit, look at the circuit diagram. Trace the path of the current round the circuit. Then connect up the components in the same sequence.
- For each circuit, say what it does when electricity passes through it.

All the way round

For a lamp to light up, it must be part of a complete circuit. There must be a complete circuit of metal all the way from the positive end of the battery to the negative end.

Metals are good **conductors** of electricity. A material which does not conduct electricity is called an **insulator**. The wires which connect components together are usually covered in plastic, which is a good insulator. If the wires touch, electricity can't flow between them because the plastic blocks its path.

There are three electrical wires here, each with a different colour of plastic insulation. Each wire is made of many strands of copper

Summary Questions

1. Explain why the filament of a light bulb is made of metal.

2. Draw a circuit diagram showing a complete circuit with a battery, lamp and switch. Label each of the three components with its name.

3. The drawing shows a complicated electrical circuit. What will happen:
   a) when Switch 1 is closed?
   b) when Switch 2 is closed?

4. Draw a circuit diagram to show how you would test different materials to see if they conduct electricity. Label the components and write a sentence or two to explain how to use the circuit.

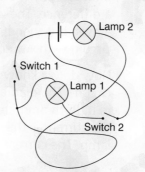

Next time you...

... plug in and switch on an electrical appliance, think about the complete electric circuit you have created, all the way to the power station where the electricity is generated.

KEY WORDS

circuit diagram
component
conductor
insulator

Electric Current

> ▶▶ How can we picture the flow of electric current in a circuit?
>
> ▶▶ How can we measure electric current?
>
> ▶▶ How can we change the current in a circuit?

What is electricity?

Usually, we can't see electricity. We can't look at a wire and see the electricity in it. However, in a thunderstorm the flash is electricity!

When a circuit is completed, an **electric current** flows in the wires. A flash of lightning is a giant electric current, zapping down from the clouds to the ground below.

A lightning flash – a huge electric current

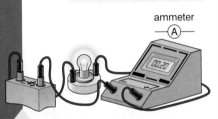

This circuit has a lamp and an ammeter in it; both show that there is a current flowing

Detecting current

Here are two ways to show if there is a current flowing in a circuit:

- Connect a lamp in the circuit. If the lamp lights up, there is a current flowing through it. The bigger the current, the more brightly it shines.
- Connect an **ammeter** in the circuit. This will measure the electric current. It will tell you how many **amps** of current are flowing.

ⓐ The electric circuit in the picture has a lamp and an ammeter in it. If the current gets less:

a) how will the brightness of the lamp change?

b) how will the reading on the ammeter change?

Around and around

We say that electric current 'flows around' a circuit. We can think of the current in a wire as being like water flowing in a pipe.

In a log flume like this, the water flows all the way round, like electric current in a circuit

Have you ever been for a ride in a fairground like the one in the photograph? It's like a ride on a fast-flowing river, but there is a difference. The water that carries the boat flows downhill like a river, but then it is pumped back uphill again. The ride is like a circuit, with water flowing all the way round it.

How do we know that electric current flows all the way round a circuit? The diagram shows a circuit with a lamp and *two* ammeters in it. Both ammeters show the same reading. This shows us that the current is the same before and after it has passed through the lamp. The current doesn't get used up in the lamp.

In this circuit, both ammeters show the same value of current

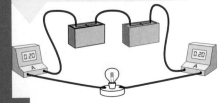

ⓑ Draw a circuit diagram for the circuit with two ammeters.

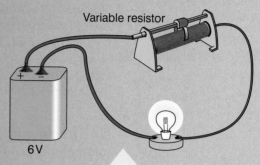

Controlling current

If you want to change the electric current in a circuit, you can use a **variable resistor**. This can make a lamp brighter or dimmer. It can make a motor turn faster or slower.

- Build a circuit to control the brightness of a lamp.
- What happens as you adjust the variable resistor?
- Add an ammeter to the circuit.
- How does the reading change as the lamp gets brighter?

Variable resistor

6 V

This circuit includes a variable resistor to control the current flowing

Electrons on the move

In a metal, there are enormous numbers of tiny particles called **electrons**. They are amazingly tiny, even smaller than the atoms of which everything is made.

Electrons move through the metal when a current flows. They flow out of one end of a battery, through the different components of the circuit, and into the other end of the battery. We need a complete circuit of metal so that electrons can flow all the way round.

Did You Know?

A million, million, million, million, million electrons weigh a kilogram.

Summary Questions

① a) What instrument do we use to measure an electric current?
  b) Which particles carry an electric current around a circuit?
  c) Which component do we use to control the brightness of a lamp in a circuit?

② The diagram shows a circuit with two lamps and an ammeter connected to a battery.

  a) Copy the diagram and label the components with their names.
  b) Add arrows to show how the current flows around the circuit.
  c) Explain why there must be two wires connected to an ammeter.

The ammeter shows that a current of 1 A flows into the first lamp. (The letter A stands for 'amps'.)

  d) How much current flows into the second lamp? Explain how you know. How could you alter the circuit to check your answer?

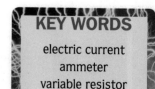

KEY WORDS

electric current
ammeter
variable resistor
electron

Cells and Batteries

▸▸ What does 'voltage' mean?

▸▸ What happens when we use more than one cell in a circuit?

▸▸ How does the voltage affect the current in a circuit?

Making the push

Every electric circuit needs something to push the electric current round. We use batteries to make our torches, MP3 and MP4 players and mobile phones work.

 Name some other devices which need batteries to make them operate.

A battery is usually labelled with the **voltage** it provides. In the photograph, the batteries have different voltages. A battery might provide a voltage of 1.5 V, 3 V or 4.5 V (the letter V stands for volts).

A **cell** is the simplest form of battery. A single cell usually provides 1.5 V. In science, when we say 'a **battery**' we mean two or more cells joined together.

The right voltage

Many devices need more than one cell to make them work. Take care! The cells must be put in end-to-end, and the right way round. Then their voltages will add up. The diagram shows how the cells should be arranged, with the negative (-) end of one next to the positive (+) end of the next one.

Each of these batteries is marked with the voltage it provides – look for the number with a V after it

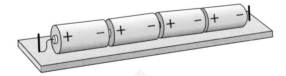

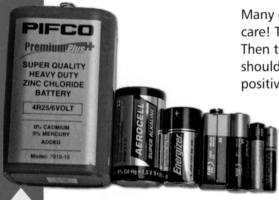

Four 1.5 V cells make a 6 V battery – if you connect them the right way round!

Botanical batteries

You can make a cell from many types of fruit and veg.

- Just stick two pieces of metal into an orange or a lemon – or even a potato.
- Now connect up to a lamp.
- Does the lamp glow as brightly as with a 1.5 V cell?
- Can you estimate the voltage provided by your fruit battery?
- Now try to make the lamp brighter by connecting two fruit batteries together.

activity

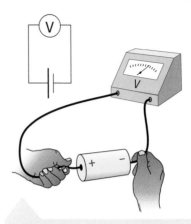

More volts, more push

What does the word 'voltage' mean? The voltage of a cell or battery tells us how much 'push' it can give to make a current flow round a circuit.

A bigger voltage means more push, so that a bigger current flows. Then the lamp in the circuit will glow more brightly. We can use a **voltmeter** to check the voltage provided by a cell or battery.

Brighter bulbs

- Connect a single cell to a lamp. Note how bright it is.
- Now use two cells to give the current more of a push.
- Make a prediction: what will happen to the lamp? Can you explain this?
- Use an ammeter to find out how the current changes when you increase the voltage in the circuit.

1 The voltage of a cell tells us how much push it gives to make an electric current flow in a circuit.

a) What instrument do we use to measure the voltage of a cell?

b) What is the unit of voltage?

c) What voltage is provided by two 1.5 V cells joined together?

d) How many 1.5 V cells are needed to give 9 V?

2 Look at the diagram. Explain why these four cells only give 3 V.

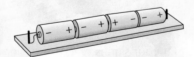

3 Find out:

a) Why is it important to recycle batteries, rather than throwing them away with other rubbish?

b) Are there any battery recycling facilities in your neighbourhood?

Summary Questions

KEY WORDS

voltage
cell
battery
voltmeter

Shocking Stuff!

>> What are the dangers of electricity?

>> How can we use electricity safely?

Modern kitchens make great use of mains electricity

Danger – high voltage

You can handle batteries safely. That's because their voltage is low so they won't give you a shock. **Mains electricity** is much more dangerous.

In the UK, the mains voltage is 230 V. We need such a high voltage to make powerful appliances work. Examples include washing machines, cookers and televisions.

(a) How many electrical appliances can you see in the photograph?

230 V is quite enough to push a big current through you. That's what we mean by an electric shock. However people aren't made of metal! Our bodies are about 70% water and water is quite a good conductor of electricity. So electricity can flow through us.

- If the current is small, it can cause your muscles to contract, throwing you across the room.
- A bigger current can burn your skin.
- A really big current can kill, by stopping the regular beating of your heart.

A second shock may be used to re-start a patient's heart. You may have seen this done in TV dramas – in real life, it is rarely successful.

Did You Know?

You can buy special waterproof radios to listen to in the shower – they run on batteries, of course.

This paramedic is trying to resuscitate a patient; you can see the two large electrical connectors being pressed onto the man's chest

That was "Singing In The Rain".... and now, Handel's Water Music.

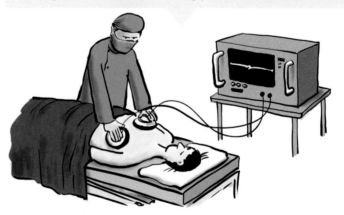

If your skin is wet, it is easier for a current to get into your body. That's why people say that 'water and electricity don't mix'. It also explains why we don't have electric sockets or switches in a bathroom.

(b) How are switches made safe in a bathroom?

Heating up

When an electric current flows in a wire, the wire may start to warm up. It may get too hot to touch and may even start glowing brightly. We make use of this in electric heaters, cookers and light bulbs.

What happens if the current flowing gets too big? An appliance may overheat and burst into flames. That can lead to a serious fire, as the photograph shows.

Staying safe

We can use the **heating effect** of an electric current to protect ourselves. An electric plug usually contains a **fuse**, which is type of safety device.

- Inside the fuse is a very thin piece of wire. The current flows through this wire.
- If the current gets too big, the wire gets hot and melts. This breaks the circuit and the current stops flowing.

When a fuse 'blows', it can easily be replaced. However the electrical fault must be sorted out first.

Some houses have fuses to protect the wiring which carries electricity around the house. Others use **trip switches** to do the same job. You can easily reset these once the fault has been fixed.

This fire in a camper van was caused by an electrical fault which caused cables to heat up and catch fire

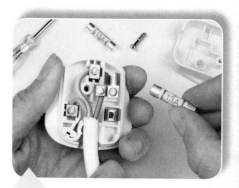

If the fuse in a plug blows, it is easy to replace

Fuse action

- Watch your teacher connect a power supply to a lamp and a length of fuse wire. The lamp will show you when a current is flowing.
- Watch what happens as the power supply is turned up.
- The greater the voltage, the greater the current that flows.
- With a high current, the fuse wire will glow and then melt.

---

1. Name two devices we use to prevent dangerous currents flowing.
2. a) What is the voltage of mains electricity in the UK?
   b) Why are you more likely to get a shock from mains electricity than from a battery?
3. List some electrical devices which use the heating effect of a current:
   a) to produce heat
   b) to produce light.
4. Usually, you can't see the fuse wire inside a fuse. How could you test a fuse to see if it had 'blown'?

Summary Questions

KEY WORDS

mains electricity
heating effect
fuse
trip switch

Series and Parallel

What is the difference between series and parallel circuits?

How does electric current flow in series and parallel circuits?

One thing after another

In the circuits you have studied so far, the components are all connected together to form a simple loop. We say that they are connected **in series**. In a series circuit:

- the components are connected end-to-end,
- the current flows from one end of the battery, through each component in turn, and back to the other end of the battery,
- the current is the same all the way round the circuit.

Two designers checking some complicated computer circuits

a Look at the circuit in the picture. Trace around it with your finger, starting from the positive (+) end of the battery. Are the components connected in series? How can you tell? Draw a diagram to represent this circuit.

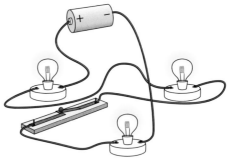

activity

Measuring current

- Set up a series circuit using the equipment provided.
- Include an ammeter to measure the current coming out of the battery or low voltage power supply.
- Move the ammeter to other points around the circuit.
- Does it always give the same reading?
- Add an extra lamp to your circuit, and repeat your measurements.
- Do you find the same result as before?

Brighter lights

If you connect two lamps in series, they will both light up. However, they won't be very bright. A single lamp would be brighter. If you switch one lamp off, they both go off.

This wouldn't be much good in a house. Usually, we want to be able to switch one light on or off without affecting all the others.

In fact, the lights in a house are connected in a different way. They are connected **in parallel**. Look at the picture on the next page. The lamps are connected side-by-side, not end-to-end. Now the current has two paths it can flow along.

- The current flows from the positive end of the battery.
- Then it divides, so that some flows through one lamp and some through the other.
- After the lamps, it joins up again and flows back to the battery.

These two lamps are connected in parallel

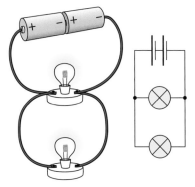

Switches in control

- Set up a circuit like the one in the picture, with two lamps in parallel.
- Add a switch so that you can switch one lamp on and off, while the other stays on.
- Where should you place a switch which will turn both lamps on and off at the same time?
- Use an ammeter to measure the current at different points in this circuit.
- What do you find?

b If each lamp in the picture has 2 A flowing through it, how much current flows from the battery?

Feeling the push

When we connect the two lamps in parallel, they both shine brightly. This is because each one feels the full push of the battery. A voltmeter will show you that each lamp has the same voltage across it.

1 What types of circuit are being described here?
  a) components connected end-to-end,
  b) components connected side-by-side,
  c) current splits as it goes round the circuit,
  d) current is the same all the way round.

2 a) Draw a circuit diagram showing several different components connected together. Include some switches in your circuit. At least two components must be connected in parallel.
  b) Explain what happens when the circuit is complete and the switches are closed. What happens if one or more of the switches is opened?
  c) Draw arrows to show how the current flows round the circuit.

3 Look at the two circuits in the picture. What current flows at the points marked A, B, C and D?

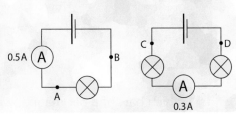

Summary Questions

KEY WORDS

in series
in parallel

Magnetic Forces

>> Which materials are magnetic?

>> Why do we talk about magnetic poles?

>> How do magnetic poles affect each other?

The magnet attracts the clips – and then the clips attract more clips

Feel the pull

We can use a magnet to attract a piece of iron or steel, even without touching it. That seems odd – usually we have to touch things to make them move. Iron and steel are **magnetic materials** – they are attracted by magnets and we can make them into magnets.

 The table shows that some metals are magnetic and some are not. How would you test a piece of metal to see if it was magnetic?

Magnetic materials	Non-magnetic materials
iron	aluminium
steel (most types)	copper
nickel	silver
cobalt	plastic (most types)
magnetic iron oxide	wood

North and south

A bar magnet is a **permanent magnet**. It keeps its magnetism for a long time. The ends of a bar magnet have a stronger pull than the middle. The ends, where the pull is strongest, are called the magnet's **poles**.

- One end of a bar magnet is usually marked to show its north pole (N).
- The other end is its south pole (S).

activity

Pole position

- Hold a bar magnet in one hand. With your other hand, move a paperclip around the magnet. Can you find its poles?

- Place two bar magnets close to each other, end-to-end, but not touching.

- How must you arrange them:

a) so that they attract each other?

b) so that they repel each other?

 Which part of a bar magnet will not attract a piece of iron?

Rules of attraction

If we bring two N (north) poles close together, they repel each other. So do two S (south) poles. If we want two magnets to attract each other, we must bring a N pole close to a S pole.

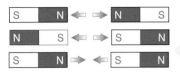

The red arrows show the forces between two magnets

The diagrams show the **forces** between two bar magnets.

We can summarise the rules of magnetic attraction and repulsion like this:

- **Like poles repel, opposite poles attract.**

Pole puzzle

- Puzzle your friends. Hide two or three bar magnets under some card. Can they work out where the magnets are without looking under the card?
- Can you design a magnetic game or toy?

Did You Know?

A magnet's poles are not exactly at its ends – they are a small distance inside it.

Would you look better with a moustache of iron filings?

Summary Questions

1. What do we call the regions of a magnet where their attraction is strongest?

2. The diagram shows two bar magnets, placed side-by-side.
   a) Explain why the magnets attract each other.
   b) Copy the diagram and add arrows to show the forces acting on the magnets.
   c) How could the magnets be arranged side-by-side so that they *repelled* each other? Draw a diagram to show your answer, with force arrows.

3. Look at the photograph of the magnet on the opposite page. How can you tell that the clips have become magnets when they have become attached to the permanent magnet?

4. Suppose your teacher gave you several pieces of steel. They all look identical, but some are magnets, others are not.

   How can you test the pieces of metal to discover which are magnets? You are not allowed to use any other materials.

KEY WORDS

magnetic material
permanent magnet
pole
force

Making Magnets

>> How can we make and destroy permanent magnets?

>> How can we best test the strength of a magnet?

Getting magnetised

If you have one magnet, you can make another. Here's how.

- Find a piece of iron or steel – a nail will do.
- Hold a bar magnet by the S pole.
- Stroke the N pole along the nail, from one end to the other.
- Make sure you always stroke the nail in the same direction, using the same pole of the magnet.

When you have stroked the nail a few times, it will be a magnet – we say that it is **magnetised**. Stroke it some more and it will be a stronger magnet.

In the picture, you can see that the end where you start stroking the nail has become a N pole. The other end is a S pole.

a What would happen if you stroked the nail with the S pole of the bar magnet?

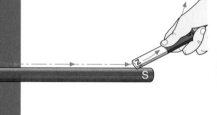

Stroke the nail from one end to the other to turn it into a magnet

activity

Make your own magnet

- Magnetise a steel needle or an iron nail, as described above.
- How can you test your magnet to show that it is magnetised?
- How can you tell which end is its N pole?

Trial of strength

How can we tell how strong a magnet is? There are several ways to **measure** the strength of a magnet. The diagrams show some different methods.

Is this possible?

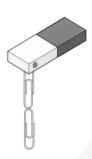

Hang paper clips end-to-end from the magnet

Hang one paper clip from the magnet, then add weights

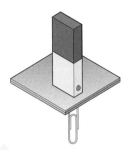

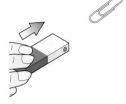

Put layers of cardboard between the magnet and the paper clip

Push the magnet towards a paper clip and watch when the clip moves

Move the magnet slowly towards the compass and watch the needle change direction

Testing magnets

activity

- Use one or more of the methods shown to measure the strength of a magnet.
- If you have two magnets, can you discover which is the stronger?

b For each method, say what quantity you would measure to find the strength of the magnet. (For example, in the first method, you could count the number of paper clips.)

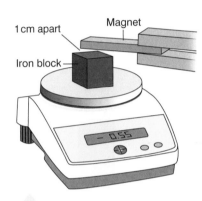

1 cm apart

Magnet

Iron block

− 0.55

Use a balance to measure the force of the magnet pulling upwards on the iron block.

Evaluating the strength tests

There are several ways we can measure the strength of a magnet. Which is the best? When we think about which method is best, we say that we are **evaluating** them.

For example, you might be trying to compare two magnets. You find that they can each hold three paper clips. This method is not very **sensitive**. However, you might find that one could lift a 30 g weight, but the other could only lift 25 g. Now you could tell that the first magnet was stronger than the second.

Yes, it's strong!

Summary Questions

1 What one word means 'made into a magnet'?

2 John says, 'You can make a needle into a magnet by rubbing it with a magnet.' This is not a very clear description of how to do this. Write better instructions for magnetising a needle.

3 Imagine that your teacher gives you two magnets and asks you to discover which is the stronger. Describe how you would go about this.

KEY WORDS

magnetise
measure
evaluate
sensitive

A Field of Force

How can we represent the field around a magnet?

How do magnets behave in the Earth's magnetic field?

Visualising the invisible

A magnet affects things around it without having to touch them. We say that there is a **magnetic field** all round the magnet.

- The field is strong when you are close to the magnet and weak when you are further away.
- The field is strongest close to the poles.

The photograph shows how we can 'see' the magnetic field around a bar magnet, using iron filings. The tiny pieces of iron line up end-to-end, showing how the field comes out of one pole and goes round to the other.

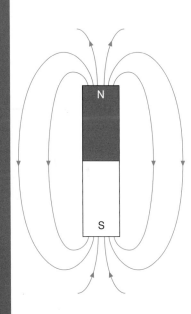

Iron filings line up to show the shape of the magnet's field

The diagram shows more clearly the shape of a bar magnet's field. We draw **lines of force** coming out of the magnet's N pole and going into its S pole. The lines are a way of showing the shape and strength of the field.

a How can you tell from the diagram that the field is strongest close to the magnet's poles?

It's all a plot

activity

- Place small plotting compasses at different points around a magnet. They will show up the shape of the magnetic field.

- Line two magnets up with their N poles 20 cm apart. Move a compass from one N pole to the other.

Compass needles turn to show the direction of the lines of force

One giant magnet

You can use a **compass** to find out which direction is North. The needle of the compass is a permanent magnet. It can turn freely, and it points North–South.

This tells us that the Earth behaves like a giant magnet. It has a magnetic field like that of a bar magnet, with lines of force coming out of one pole and going into the other. There is a lot of iron in the Earth's core and this makes it magnetic.

b The Moon's magnetic field is much weaker than the Earth's. Suggest why this is.

This man is using a compass to check where he is going (perhaps he should look at the signpost behind him)

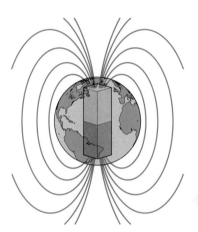

The Earth has a magnetic field all round it, as if there were a giant bar magnet inside

Did You Know?

When European scientists first made compasses, they thought that a compass needle pointed North because it was attracted to the Pole Star. Then they noticed that the needle tipped slightly downwards, into the Earth.

Summary Questions

1 Name two ways to show up a magnetic field.

2 When we draw lines of force to represent a magnetic field:
   a) Where do the lines start and finish?
   b) How can you tell where the field is strong?

3 Look at the diagram. It shows the field lines around two magnets. They are repelling each other.

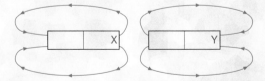

   a) How can you tell from the diagram that they are repelling each other?
   b) Is pole X a N pole or a S pole? How can you tell?
   c) Is pole Y a N pole or a S pole? How can you tell?

4 If you used a microscope to look at a magnet, would you be able to see the lines of force? Explain your answer.

KEY WORDS

magnetic field
lines of force
compass

Electromagnets

>> How can we make an electrically operated magnet?

>> How can we increase the strength of an electromagnet?

Switch on, switch off

The photograph shows a giant **electromagnet**. It is at work at a scrap metal yard. The crane driver can switch on the magnet; it attracts pieces of scrap iron and they can be moved around the yard. When they are in the right place, the driver switches off the magnet and they fall to the ground.

An electromagnet is only magnetic when an electric current flows through it. When the current stops, the electromagnet stops attracting things.

> **a** What is the difference between a permanent magnet and an electromagnet?

Inside an electromagnet

Every electromagnet contains a coil of wire, sometimes called a **solenoid**. When a current flows in the coil, a magnetic field appears around it. One end of the coil is the N pole, the other end is the S pole. When the current stops, the magnetic field disappears. Most electromagnets have an iron **core** inside them. This makes the magnetic field stronger.

A coil, a switch, a current – it's an electromagnet, lifting scrap metal

activity

Making an electromagnet

The diagram shows a model electromagnet which you can make and test in the lab.

- Wind a coil of wire around a cardboard tube.
- Connect to a low voltage power supply and switch on.
- Will your electromagnet attract a paper clip?
- Try again, with an iron nail as a core.

Stronger electromagnets

A scrap yard electromagnet is very strong.

- Its coil is made of many turns of wire. The more turns of wire, the stronger the magnetic field of the electromagnet.
- A big current flows through it. The bigger the current, the stronger the magnetic field.
- It has an iron core. A core makes any electromagnet much stronger.

It's the current in the wire that makes the magnetic field. A coil is a clever way of concentrating the magnetic field in a small space.

I guess I wired it up the wrong way round!

b Draw two electromagnets; one must be much stronger than the other. Indicate the features that make it stronger.

activity

Testing an electromagnet

- You have made an electromagnet. Now you have to test it.
- How can you show that it is stronger if you increase the current flowing through it? Can you show that twice the current gives twice the force of attraction?
- How can you show that an electromagnet is stronger if it has more turns of wire, or an iron core?

science @ work

The patient in the photo is having a magnetic resonance scan. She is lying beneath a giant electromagnet – but she won't feel a thing.

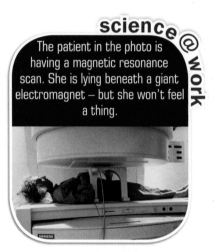

❶ Give another name for the coil of an electromagnet.

❷ Why is iron a suitable material to use as the core of an electromagnet?

❸ Would copper wire be a suitable material:
  a) for the coil of an electromagnet?
  b) for the core of an electromagnet?
  Explain your answers.

❹ An electromagnet with twice the number of coils produces twice the attractive force. Describe in detail how you could test this idea. (Remember, you must devise a fair test.)

KEY WORDS

electromagnet
solenoid
core

Electromagnets at Work

> ▶▶ What is the pattern of the magnetic field around an electromagnet?
>
> ▶▶ How do we use electromagnets?

From north to south

An electromagnet is like a bar magnet, except that you can switch it on and off. The photograph shows how iron filings can be used to show up its magnetic field.

- There is a N pole at one end and a S pole at the other.
- Its magnetic field lines come out of the N pole and go round to the S pole.

The diagram shows that, with twice the current, the field is twice as strong. We say that the field is directly proportional to the current.

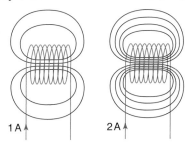

1 A   2 A

The two electromagnets are identical, but the one on the right has twice the current flowing through it

A current is flowing through the black electromagnet; iron filings line up to show the magnetic field

ⓐ How can you tell from the diagram that the magnetic field is twice as strong when twice the current flows?

ⓑ How could you find out if there is a magnetic field around a straight piece of wire when a current flows through it?

The field of an electromagnet

activity

- Set up an electromagnet. Place plotting compasses around it and switch on.
- Do the compasses line up in the field? Do you see the same pattern as with a bar magnet?

Electromagnets everywhere

Ring ring! Ding dong! Doorbells and chimes use electromagnets. The hammer that strikes the gong of a doorbell is operated by an electromagnet.

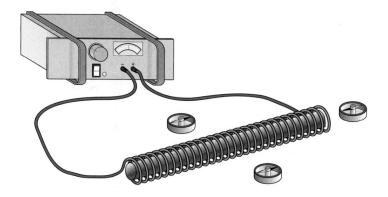

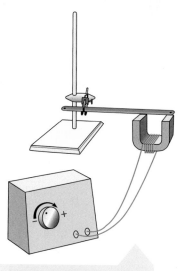

An electromagnetic clicker

The diagram shows a simple **device** that illustrates how this works.

● A coil of wire is wound on a C-shaped iron core.
● A flexible steel strip is fixed above the core.
● When a current flows through the coil, it becomes magnetised and attracts the steel strip. Click! The strip bends and hits the core.
● Switch off the current and the strip bends back again.

An electric bell is cleverly designed so that the bendy strip moves back and forth automatically, so that you get a continuous ringing, not just a click.

An electric bell – look for the electromagnet, the bendy strip, the hammer and the gong

How it works

● Examine an electric bell and discover how it works.

● Investigate a relay – a type of electromagnetic switch.

● Find some old headphones that no longer work. Dismantle them carefully – you will find a coil of wire (an electromagnet) and a strong permanent magnet.

science@work

In a car accident, you might get a splinter of metal in your eye. The surgeon will pull it out using an electromagnet.

Summary Questions

❶ Make a list of some devices which make use of electromagnets.

❷ The strength of an electromagnet is 'directly proportional' to the number of turns of wire. Explain what this means.

❸ What do you think would happen to the magnetic field of an electromagnet if you made the current flow in the opposite direction? How would you test your idea?

❹ Trip switches are often used instead of fuses (page 113). There is an electromagnet in a trip switch – find out how it works.

KEY WORDS

device

▼ Question 1 (level 4)

The circuit shown below contains one switch and three bulbs.

a What other component is there in the circuit?
[1]

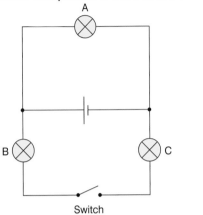

b Say whether each bulb is on or off. [3]

▼ Question 2 (level 5)

Jo is investigating the magnetic field of a bar magnet. She places a small plotting compass near one of its poles, B, as shown in the diagram.

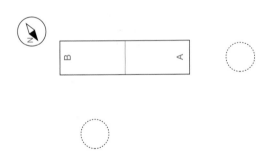

a Is Pole B a north magnetic pole or a south magnetic pole? Explain how you know. [2]

b The diagram shows two other positions in which Jo placed the compass. Copy the diagram and mark on it the direction of the plotting compass needle in these two positions. [2]

Jo hangs the magnet up using a length of string. The magnet is free to turn around.

c In which direction does pole A point? Explain how you know. [2]

▼ Question 3 (level 5)

Bob set up the circuit shown. He had three bulbs to test. Each gave a different reading on the ammeter.

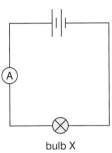

bulb X

Bulb X	Ammeter reading = 0.4 A
Bulb Y	Ammeter reading = 0.2 A
Bulb Z	Ammeter reading = 0.6 A

a Which bulb lets through the least electrical current? [1]

b Which bulb has the least resistance? [1]

Bob then set up a circuit with two of the bulbs connected to the same battery as before.

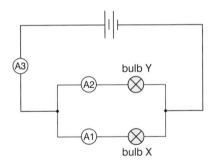

c Are the bulbs connected in series or in parallel? [1]

d What readings would you expect to see on ammeters A1 and A2? [2]

e What reading would you expect to see on ammeter A3? [1]

How Science Works

Mel's teacher asked her to test some batteries. Mel connected each battery in turn to a bulb. She used a light meter to measure the brightness of the bulb.

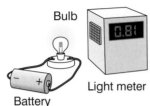

The table shows her results.

Battery	Reading on light meter
X	0.70
Y	0.95
Z	1.00

a How many batteries did Mel test? [1]

b Which battery made the bulb brightest? [1]

c Draw a bar chart to represent Mel's results. [2]

d Which two things should Mel have done to make her test fair? Give the letters. [2]

    A   Use the same bulb each time.

    B   Connect each battery the same way round.

    C   Keep the distance between the lamp and the light meter the same each time.

    D   Keep the bulb on for the same time with each battery.

    E   Repeat her measurements with a different bulb.

Josh designed a way of comparing two electromagnets. He lined them up, as shown in the picture, with a gap between their ends. He hung a magnet in the gap.

Josh said, 'If one is stronger than the other, the magnet will move towards the stronger one.'

When Josh connected up his circuits, the magnet moved towards electromagnet A. Josh said, 'This shows that the current through A must be bigger than the current through B.'

a Give another reason why electromagnet A might be stronger than electromagnet B. [1]

b Suggest one way in which Josh could redesign his experiment to ensure that the two electromagnets have the same current flowing through them. [2]

Pete's teacher gave him an electrical 'black box' to investigate. Pete connected the black box to a cell. He included an ammeter in the circuit to measure the current flowing through the black box.

Pete added more cells, and recorded the current each time. The table shows his results.

Number of cells	Current flowing (A)
1	0.40
2	0.80
3	1.40
4	1.60

a When the teacher looked at Pete's results, she suggested he might have made a mistake with one of them. Which one? [1]

b Draw a graph of Pete's results. Using a ruler, draw a line through his correct results. [2]

c Use your graph to make a prediction: what current would flow if the black box was connected to 5 cells? [1]

Forces and Energy

Getting stronger

- As you are getting older, you are getting stronger. You can lift a heavier bag of school books; you can push a car when it breaks down; you can break a thicker piece of wood. The force provided by your muscles is getting bigger.

- However, as you get older, you also get wiser. You learn that you don't have to build giant muscles to do these things. It's easier if you carry those heavy books on your back. A truck can pull the broken-down car. A saw will cut through the wood. There is more to forces than just musclepower.

Could you do this?

- You already know quite a lot about forces – forces that stretch; the force of gravity (weight); the force of air resistance. In this topic, you will learn more about how forces work.

Push, pull, twist

- Here are three words to describe how a force acts:
  - A force can **push**.
  - A force can **pull**.
  - A force can **twist**.

a Take a pencil or a pen. How can you use it to demonstrate these different ways of making a force act?

This woman is a base jumper; she makes use of forces – gravity pulling her down, air resistance pushing up strongly when her parachute opens

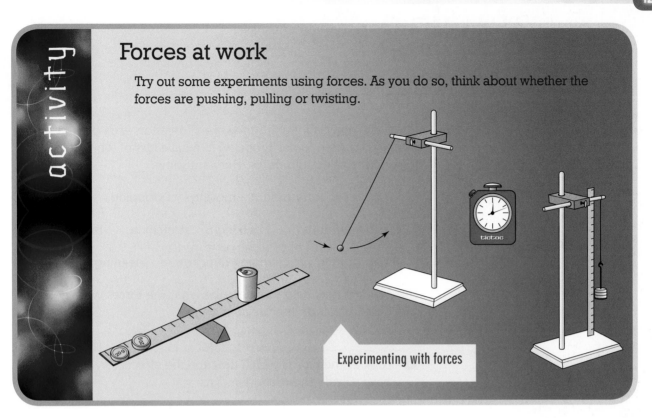

activity

Forces at work

Try out some experiments using forces. As you do so, think about whether the forces are pushing, pulling or twisting.

Experimenting with forces

Feeling the force

- Rub you hands together – you feel the force of friction. Sit on your hand – you feel your weight pressing down on your hand. Accidentally bite your lip – that hurts!
- It takes two things to produce a force – one hand rubbing on the other, your teeth biting your lip. Each one feels a force.

Did You Know?

We can't see forces, but we can certainly feel their effects. There are nerve endings all over our bodies which respond to forces. If the force gets too big, we feel pain.

b What forces are at work when you slide down the banisters (or down a water slide)?

c Re-read these two pages and make a list of the different forces mentioned.

d Look at the photograph of the woman base jumper on the **opposite** page. She opens her parachute so that she will land safely. Draw a diagram to show the forces acting on her as she falls.

e The photograph shows a boy weighing his pet cat. Explain what measurements he must make, and how he will find the cat's weight.

f Choose an area of interest. Describe some of the ways in which forces are important in your chosen area. Choose from:

sport   music and dance   cooking   transport

Measuring Forces

* What effects do forces have?
* How do we measure forces?
* How do we represent forces on a diagram?

An example of a forcemeter

What forces do

When two objects interact, they exert forces on each other. What *changes* do these forces produce? We have lots of words to describe the effect of a force. Here are some:

squashing   moving   stopping

slowing down   breaking   stretching   bending

turning   changing direction   speeding up

(These words may help you describe what the forces were doing in your experiments on the previous page.)

> **a** Look at the words which describe the effects forces can have. Divide them into two lists:
>
> **Forces changing how an object moves**
>
> **Forces changing an object's shape**
>
> Can you add some more words to the lists?

Newtonmeters

To measure a force, you use a **forcemeter**, also called a **newtonmeter**. Some forcemeters you pull to measure the force; others you push.

Inside the forcemeter is a spring, which is stretched or squashed by the force. The bigger the force, the more it stretches or squashes.

The unit of force is the **newton (N)**. To get an idea of the size of a newton, imagine holding an apple on the palm of your hand. Its weight, pressing down on your hand, is about 1 N.

activity

Measuring pushes and pulls

* Use newtonmeters to weigh some objects and to measure some other forces.

* If you use two different newtonmeters to weigh the same object, do they both give the same answer?

Representing forces

The cars in the photograph have collided. You can see that each car has exerted a force on the other car.

In the diagram, the arrows show these forces.

- The *direction* of the arrow shows the direction of the force.
- The *length* of the arrow shows the size of the force.

The label tells us two things:

- what caused the force,
- what the force acted on.

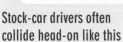

Stock-car drivers often collide head-on like this

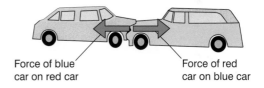

Force of blue car on red car   Force of red car on blue car

> **b** How can you tell from the diagram that the two forces were the same size? Did the two forces act in the same direction?

1 Copy these sentences and complete them by filling in the gaps:

Forces are measured using a …

The scientific unit of force is the … (symbol …).

2 Draw a diagram to show a foot kicking a ball. Show the force of the foot on the ball.

3 Jo and Al weigh a block of wood. Jo's forcemeter shows that it weighs 18 N. Al's forcemeter shows that it weighs 20 N. How would you try to decide which (if either) is correct?

4 When two magnets are brought close together, they can exert forces on each other without even touching. Can you think of any other forces which objects exert on each other without touching?

KEY WORDS

forcemeter
newtonmeter
newton (N)

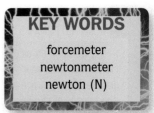

Bending, Stretching

▶▶ What causes us to have weight?

▶▶ How can we show the pattern when a force stretches a spring?

This baby is being weighed as part of a routine health check

Which is heavier?

Next time you...

... go up in an aircraft, remember that your weight has decreased – you are further from the centre of the Earth. However it will be back to normal when you touch down again.

Gravity causes weight

Weight is the name of the force which we give to the pull of the Earth's gravity on an object. Every object on or close to the Earth feels this pull, acting towards the centre of the Earth.

Because weight is a force, it is measured in newtons. You need a forcemeter with a scale in newtons to measure weight.

Earth's gravity gets weaker as you move further away from the Earth. If you went to the Moon, your weight would be much less. That's because the Moon is much smaller than the Earth and so its pull is much weaker.

a The photograph shows one way in which we use forcemeters in everyday life. Give some more examples.

activity

Comparing weights

- Try holding two objects of similar weights, one in each hand. Can you judge which is the heavier?

- Now try weighing them using forcemeters. Can you tell which is the heavier?

- Test yourself – what is the smallest *difference* in weight you can detect simply by holding an object in each hand?

Stretching springs

A forcemeter has a spring inside. The scale on the meter tells you what the force is which is pulling it.

The scale has evenly spaced markings on its scale. This tells us something important – the spring stretches evenly when the force pulling it increases at a steady rate.

You can investigate this property of a spring by measuring two things:

- the **load** on the end of the spring.
- the **extension** produced by the load.

(The extension is the *increase* in length, compared to the spring's un-stretched length.)

(b) If a spring is 6 cm long when un-stretched and 10.5 cm long when a load is added, what is its extension?

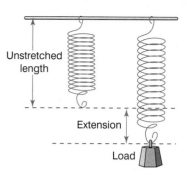

Unstretched length

Extension

Load

Investigating a spring

Previously, you may have investigated an elastic band. Now you can investigate a steel spring.

Work with great care, so that your results are as accurate as possible! When you draw a graph of your results, you will be able to see how well you have done. Wear eye protection in case the spring snaps.

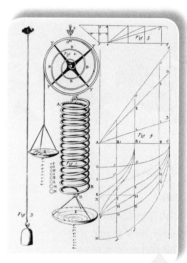

Let's get this straight …

The best way to show the pattern of results in an experiment like this is to draw a graph of extension against load. The points form a straight line. This shows that, if you increase the load in equal steps, the extension of the spring also increases in equal steps. We say that:

● The extension of the spring is *directly proportional to* the load.

This rule was first discovered by Robert Hooke in the seventeenth century. It is known as *Hooke's law*. The picture shows the spring he used. You can also see his straight-line graph.

Robert Hooke's drawing of his experiments with springs. As well as the long spring in the centre, he also investigated a spiral spring (at the top) which he wanted to use in a new type of watch

Summary Questions

1 Every object in the room around you has weight. What causes this?

2 What units do we measure weight in? Why do we use these units?

3 The table shows the results of an experiment to stretch a spring. Draw a graph to show the pattern in these results. Are there any points on the graph which you don't trust? Circle them on your graph.

Load (N)	Extension (cm)
0.0	0.0
2.0	1.4
4.0	3.0
6.0	3.8
8.0	6.2
10.0	7.5

4 You have found how the extension of a spring depends on the force stretching it. Explain how you could make a forcemeter using your spring. How would you use it to weigh a pebble?

KEY WORDS

weight
load
extension

Friction

- ▶▶ What causes friction?
- ▶▶ What factors affect friction?

If these speed skaters misjudge their speed, they will all end up in a heap

Opposing motion

Friction can be a problem, but it can also be very useful. So it is important to understand how it works.

The ice skaters in the photograph need to be able to control the friction between their blades and the ice. To keep travelling steadily along, they rely on the smoothness of the ice to reduce friction. However, to speed up or change direction, they must turn their blades sideways and push on the ice – that needs friction.

The direction of friction

If you try sliding along the floor in your socks, you will soon come to a halt. Friction slows you down; it acts in the opposite direction to your movement.

This means that, in a diagram, we can use a force arrow to show the friction on a moving object. The arrow is in the opposite direction to the direction in which the object is moving.

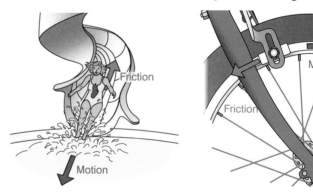

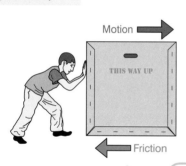

Friction always acts in the opposite direction to motion

(a) Draw a diagram to show a child pulling a toy along the ground. Show the force of friction acting on the toy.

Weight

The man pushing the box will find it easier to push if he removes some of the contents. Friction is less if the weight pushing downwards on the surface is less.

activity

Measuring friction

- You can measure friction using a forcemeter.
- Use it to pull a block along at a steady speed. The reading shows the force of friction.
- Use this method to investigate the factors which affect friction.

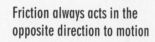

Rubbing along

Smooth surfaces produce less friction than rough ones. This gives us a clue to the cause of friction. The photograph was taken using an electron microscope. It shows the surface of a matchbox – the part where you strike a match. It is very rough, because it takes a lot of friction to generate the heat which lights the match.

So what happens when two rough surfaces slide over each other? All the rough, pointy bits get in each other's way. This makes it hard for the two surfaces to slide past each other. Wherever the two surfaces interlock, each one exerts a force on the other. All of these small forces add up to give the force of friction between the surfaces.

Lubricants, such as oil, provide a thin, slippery layer of liquid between the two surfaces. This makes it harder for them to interlock.

The striking surface of a matchbox, magnified 500 times with an electron microscope

Friction in fluids

There is friction when something moves through a liquid or a gas. It's called **drag**. For something moving through the air, it's called **air resistance**. Parachutists rely on air resistance to slow them down for a safe landing.

> **b** Draw a diagram to show the forces acting on a parachutist falling slowly to the ground.

1 Copy and complete this sentence:

Friction is a force which … motion when two … try to slide over one another.

2 What name do we give to substances which can reduce the friction between two surfaces? How does such a substance work?

3 Make a list of the different factors which affect friction. For each one, say whether it increases or decreases friction.

4 Use the picture (above) of rough surfaces to explain why friction is increased when the weight pressing down increases.

Summary Questions

KEY WORDS

friction
lubricants
drag
air resistance

Floating and Sinking

>> Why do some things float while others sink?

This professional diver is preparing to dive

Two forces act on you when you are in the water

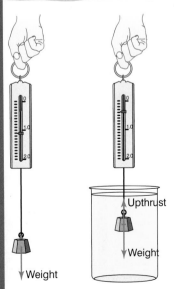

Upthrust

Weight

Weight

Diving deep

The diver in the photograph is strapping on his belt of weights. He needs to wear these heavy weights if he is to dive deep in the ocean.

If you have been in a swimming pool, you may have noticed that it is difficult to reach the bottom in the deep end. Nervous people think that, if they let go of the side, they will sink to the bottom. However it isn't like that. The water buoys you up. Like the diver, you would need to carry heavy weights if you wanted to stay down at the bottom for any length of time.

Underwater forces

Some things float in water (people, for example). Other things sink (stones, for example). We need to think about the forces at work to understand why this happens.

There are two forces acting on you in a swimming pool:

- your **weight**, the pull of the Earth's gravity, pulling you downwards,
- the **upthrust** of the water, pushing you upwards.

You can see from the diagram that these forces act in opposite directions. That's lucky – the upthrust can balance out your weight. When two equal forces act on an object, and they act in opposite directions, we say that the forces on the object are **balanced**.

- When you are underwater, the upthrust of the water is greater than your weight and you are pushed up to the surface.
- When you are floating, upthrust and weight exactly balance out, so you stay floating.

ⓐ When the diver wears weights and dives deep in the ocean, which force is bigger, his weight or the upthrust? Draw a diagram to show the forces acting on him.

activity

Measuring upthrust

Here is how to measure the upthrust on an object when it is submerged in water:

- Hang the object on a forcemeter; record its weight.
- Now submerge the object in water. The reading on the forcemeter decreases.
- The change in reading is equal to the upthrust. Explain why this tells you the upthrust.

Explaining upthrust

The photograph shows two small submersibles. They can go deep in the ocean to repair damaged oil platforms.

The submersibles are nearly spherical. They need to be this shape to withstand the high pressure of the sea.

The deeper you go in the sea, the greater the pressure of the water. The bottom of the submersible is lower in the water than the top, so the pressure on it is greater. This means that there is more force pushing upwards than pushing downwards – the difference in force is the upthrust.

These engineers are welding an underwater oil pipeline; they are protected by the spherical shape of their submersibles

Weight

Upthrust

The water provides an upward push on the submersible

Summary Questions

1. A cork floats on water. Which two forces are balanced? Draw a diagram to show these forces.

2. A stone sinks to the bottom when placed in water. Draw a diagram to show the forces on it as it falls through the water. Are the forces balanced?

3. A beaker will float in water. Add some sand and it will sink deeper in the water. Add more sand and it will eventually sink. Describe how the forces on the beaker change as the sand is added and use this to explain why it eventually sinks to the bottom.

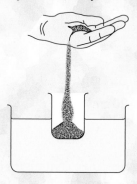

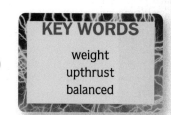

KEY WORDS

weight
upthrust
balanced

A Matter of Density

▸▸ How can we find out the density of a substance?

▸▸ How can we predict which things will float?

Heavy stuff

When an iceberg floats, most of it is hidden below the surface. That's what makes icebergs dangerous to ships – you can't see the part that is under water.

An iceberg floats because ice is lighter than water.

Most of an iceberg is under water, where you can't see it

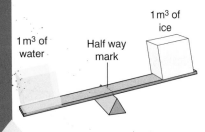

Comparing equal volumes of ice and water

What do we mean when we say that ice is 'lighter' than water? We are comparing two materials, ice and water. If we take the same volume of each – say, one cubic metre – the ice will weigh less than the water.

To use the correct scientific words, we should say that ice is *less dense* than water. In other words, the **density** of ice is less than the density of water.

Calculating density

A coin is made of metal. How can we find the density of the metal? We need to know:

- the **mass** of the metal, measured in grams (g),
- the **volume** of the metal, measured in cubic centimetres (cm³ or ml).

The mass tells us how much material (matter) it is made of. You can find it using a balance. Don't confuse mass with weight. Mass is not a force. It is an amount of matter.

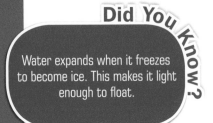

Did You Know?

Water expands when it freezes to become ice. This makes it light enough to float.

Now we calculate density:

- $$\text{density} = \frac{\text{mass}}{\text{volume}}$$

Here is an example:

- mass of stone = 46 g
- volume of stone = 20 cm³
- $$\text{density} = \frac{\text{mass}}{\text{volume}} = \frac{46\,g}{20\,cm^3} = 2.3\,g/cm^3$$

The answer is in grams per cubic centimetre (or kilograms per cubic metre). You can see that we use density as a 'fair-test' way of comparing two materials, by comparing the same volume of each, 1 cm³.

(a) Calculate the density of ice if 200 cm³ of ice has a mass of 184 g.

Measuring the volume of an object by displacing water

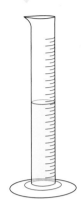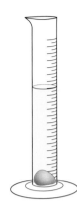

Measuring density

Find the densities of some items made of different substances. You can measure their masses using a balance.

- To find their volumes, put some water in a measuring cylinder.
- Submerge the item in the water and record by how much the volume has increased.

Density and floating

The density of water is 1.00 g/cm³. Anything with a lower density than this is lighter than water and will float. Anything with a greater density is heavier and will sink.

This allows us to predict which materials will float in water and which will sink.

Substance	Density (g/cm³)
water	1.00
sea-water	1.03
wood	0.90–1.05
ice	0.92
concrete	2.40
cork	0.24
steel	7.9
lead	11.3
gold	19.3

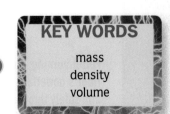

(b) Look at the table. It shows the densities of several substances. Which ones will float in water?

Summary Questions

1. Give the units of: a) mass, and b) weight.

2. You will float if you jump into a swimming pool. What does this tell you about your density?

3. Olive oil has a density of 0.92 g/cm³. If you put olive oil and water in the same bottle, which will float on top?

4. Mahogany is a dense type of wood. A 50 cm³ block has a mass of 51 g. What is its density?

5. A ship is made of steel and it floats. Does this mean that steel is less dense than water? Explain your answer.

KEY WORDS

mass
density
volume

Fuels Alight

▸▸ How can we release the energy stored in fuels?

▸▸ Which fuels store the most energy?

Cooking, heating, lighting

A barbecue is a great way to cook food. Burning charcoal produces a high temperature and the food cooks quickly. The smoke adds to the flavour. When the fire dies down, you are left with a small heap of grey ashes. The fuel has been turned into new substances.

MEGAFUEL CORPORATION

So, burning requires oxygen. Now, how can we start selling oxygen?

C.E.O

Barbecuing sausages; the hot charcoal glows brightly as it burns

Charcoal, wood and gas are all **fuels** which can be used for cooking. As they burn, they release heat and light – two forms of energy. It is the heat that cooks the food.

> **a** Name another fuel which can be used for cooking. Can you suggest a method of cooking which does not involve burning a fuel?

What is a fuel?

A fuel is a substance which must be burned to release its store of **energy**. As it burns, it combines with oxygen from the air. 'Burning' is also known as **combustion** (see page 72).

Petrol burning; the stunt rider's powerful bike carries him safely through the flames

Comparing fuels

A fuel is a store of energy. Some fuels are better than others – they store more energy.

- Devise a method of comparing fuels. How will you make it a fair test?

 Safety: Always wear eye protection when burning fuels.

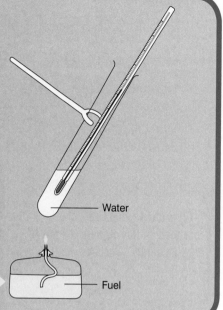
— Water

You can burn a fuel and use the energy it releases to heat some water

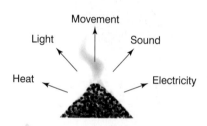
— Fuel

Fuels are useful

We use many different fuels, but we aren't always aware of them.

- A candle is made of wax. Wax is a fuel and when it burns, it gives out heat and light.
- In a central heating system, the boiler uses coal, oil or gas as its fuel. The burning fuel heats water.
- Cars run on petrol or diesel. You can't see it burning, but this is what is going on in the engine. The energy released gives us movement.

Is electricity a fuel? Not really. You don't burn electricity. Usually, fuel has been burned at the power station to make the electricity.

Light | Movement | Sound
Heat | Electricity

Fuels are stores of energy which we can burn to give us other forms of energy

b Sound is another form of energy. Give an example of a fuel burning and producing sound.

Summary Questions

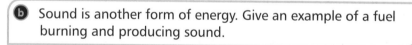

1 List all the fuels mentioned on these two pages. Add any others you know of.

2 Wood is a store of energy. How would you convince someone of this?

3 Find out what fuels are used in these:

camping stoves   aircraft   hot air balloons

space rockets   lawnmowers

KEY WORDS
fuel
energy
combustion

Burning the Past, Wrecking the Future?

Fossil fuels

Coal, oil and gas are described as **fossil fuels**. Coal is mined from under the ground. Oil and gas are extracted from wells, often out at sea. Then they are piped ashore, or carried in giant tanker ships.

▶▶ How are fossil fuels formed?

▶▶ What problems are caused when we use fossil fuels?

Oil is often found under the seabed; at this oil well, the gas is being burned – nobody wants it

This power station uses a trainload of coal every hour

Fuels by the tonne

Most of our electricity is generated in **power stations** where fossil fuels are burned. Our cars and other forms of transport also rely on fossil fuels.

In one year, the average UK citizen uses:

- 1 tonne of fossil fuels for generating electricity
- 1 tonne for transport
- 1 tonne for heating and cooking
- 1 tonne for the industry which keeps us working

That's 4 tonnes a year, for each person!

a How do cars rely on fossil fuels?

Did You Know?

Each year, we use fossil fuels which took one million years to lay down.

Fossil fuel history

Where do fossil fuels come from? They formed roughly 300 million years ago. At that time, Britain was very different. The climate was warm and the land was swampy. Giant ferns and mosses grew, and reptiles and amphibians roamed the land. There were no flowering plants or mammals.

Plants which died and fell into the swamps did not rot away completely. First they became peat; then they were gradually compressed so that they turned into underground layers of coal.

Oil and gas formed in a similar way, from plant and animal remains falling to the seabed.

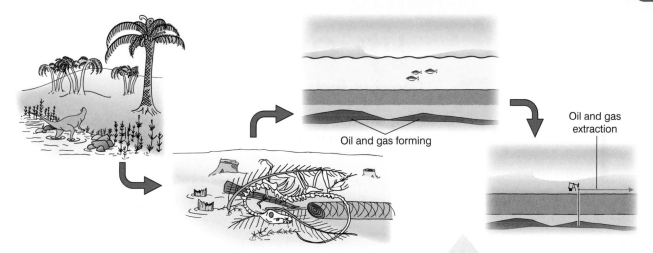

Oil and gas forming

Oil and gas extraction

Going up in smoke

Fossil fuels contain carbon. When they burn, carbon dioxide forms and that's where the trouble starts (see page 74).

Carbon dioxide is described as a **greenhouse gas**. In the atmosphere, it acts like a blanket around the Earth, warming it up. It's good to have some carbon dioxide in the atmosphere, but if we keep producing more and more, the result will be rapid **climate change**. We won't have time to adapt to hotter summers, changing patterns of rain and wind and rising sea levels.

The danger is that we will go back to the days when fossil fuels were formed – hot and swampy, with no mammals – including no humans!

Oil and gas are trapped by layers of rock which form on top

CARBON 11 TONNES

activity

Carbon footprint

The 4 tonnes of fossil fuels which we each use every year turns into about 11 tonnes of carbon dioxide in the atmosphere. That's called our carbon footprint. People who are concerned about climate change encourage us to reduce our carbon footprint.

Make an estimate of your own carbon footprint and find out how you can reduce it.

Summary Questions

1. List three fossil fuels.

2. List some everyday activities which make use of fossil fuels. For each one, suggest how we can do it without using fossil fuels.

3. When fossil fuels are burned, waste gases are produced which result in acid rain. This can poison rivers and the soil, killing plants and fishes. Find out how it is possible to cut the production of these harmful gases.

KEY WORDS

fossil fuel
power station
greenhouse gas
climate change

Renewables – Cleaning up our Act

Growing fuels

Many farmers are now in the business of growing fuels. They have done this in the past – for example, when they grew oats to feed cart horses. However you can't run a car on oats.

An important new crop is **biomass**. Fast growing trees can produce useful wood within a few years and this is burned in power stations to generate electricity.

Soya beans, maize and sugar cane are grown to make **biodiesel**. This is used in cars, buses and lorries.

Come again

If farmers are careful with their land, they can grow crop after crop, year after year. The Sun shines, their plants grow and store energy and eventually a motorist or a power station makes use of that energy.

Biofuels are described as **renewable**, because we can always grow new plants to replace the ones we use.

Fossil fuels are **non-renewable**. Once we have dug them up and used them, we can never get them back again – unless we wait for a few hundred million years!

> **How can we produce electricity without harming the environment?**
>
> **Can we manage without fossil fuels?**

This factory in Thailand converts cassava roots into biofuel

Great Debates

Not everyone thinks that biofuels are a good thing. Farmers in third-world countries may benefit, as they can sell their crops for fuel rather than food. However, other people may suffer as the price of these crops increases.

> a   If you feed oats to a horse, is that a renewable energy supply? Explain your answer.

Biodiesel can be used in conventional cars

Renewable electricity

It would be good if we could generate all of our electricity from renewable sources. Here are some ways of doing this:

- **Hydroelectric power:** Water is stored by a dam. As it flows out, it turns a turbine which turns a generator.
- **Wind power:** The wind turns a turbine and generator.

- **Wave power:** Waves on the sea have energy. This can be used to generate electricity.
- **Solar power:** Sunlight falls on a solar cell to produce electricity.

All of these are renewable sources of electricity. As long as the Sun shines, we can expect the wind to blow, rain to fall and waves to appear on the sea.

Testing renewable electricity

- Try out some model systems which generate electricity from sunlight, from moving water and from the wind.
- For each, find out an example of where it is used in practice.

Is this sustainable?

There is a problem with these energy supplies. What happens when the Sun goes in, or there is no wind or waves? We might be left without electricity. At present, the electricity companies keep some fossil fuel power stations on standby, ready to fill the gap.

We need to make use of **sustainable** energy supplies. This means that we should not use up resources or damage the environment; that would make it difficult for future generations to enjoy comfortable lives.

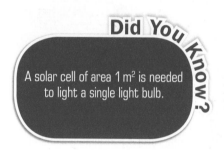

Did You Know?

A solar cell of area 1 m² is needed to light a single light bulb.

b Explain why solar power is sustainable, but petrol-powered cars are not.

Summary Questions

1 List all of the renewable energy resources mentioned on these two pages.

2 A farmer has a field of fast growing willow trees. Explain how these could be used to generate electricity. Is this a sustainable source of energy?

3 Most spacecraft have solar cells to generate electricity from sunlight. Why is this a good choice?

4 The government would like more of the UK's electricity to come from renewable sources. Find out:
- how much of electricity comes from renewable sources at present
- what these sources are
- how much renewable electricity we can expect to use in the future.

KEY WORDS

biomass
biodiesel
renewable
non-renewable
sustainable

Making More of Energy

>> How can we waste less energy?

>> Why should we try to save energy?

The walls of this new house are covered in blocks called 'Tyvek'; the roofers also use insulating material before they put the tiles on

Better buildings, by order!

When new homes are built, the builder must be sure that they are constructed to a high standard. In particular, they must be well-insulated.

In the photograph, you can see the thick panels of insulating material which cover the walls of the house being built. These will help to keep the owners warm in winter and cool in summer.

Keeping a check

In the past, houses were built to a much lower standard. Walls were thinner, windows were single glazed, and lofts had little **insulation**. In cold weather, lots of energy escaped to the surroundings.

Today, things are different. Walls, floors and the roof must be insulated to a high standard. Windows must be double glazed.

There are strict rules which builders must meet. An inspector from the council comes round to check that the rules are followed.

ⓐ Think about your school building. Are the walls well-insulated? Are the windows double glazed?

activity

Keeping heat in

Here is one way to test some insulating material:

- Carefully place a metal block in a beaker of water and heat the water to boiling point.

- Carefully take the block out and fit it with a thermometer or temperature probe.

- Surround it with insulating material.

- Record how the temperature drops.

- Repeat with the same block, without insulation.

- How could you find which insulating material is best at keeping heat in?

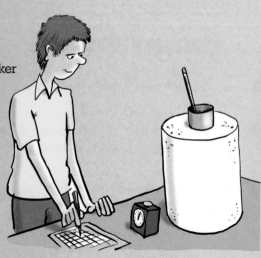

⚠ **Safety:** Take care with boiling water. The metal block will stay hot after removing from water!

Saving energy, saving money ...

In the UK, winters can be cold. We heat our homes to keep them warm. We don't want heat energy to escape through the walls, windows and roof. That would be a waste of energy and a waste of money. It costs a lot to keep a house or flat warm.

If a house is poorly insulated, the owner can add more insulation, or fit double glazing. This costs money, too, but it is worth it in the long run.

There is another way to save energy and money. Turn down the **thermostat** (the temperature controller) so that the house is a little cooler. If you wear warm clothes, you won't notice the difference.

... saving the planet

If we use less energy to heat our homes, we won't just save money. We will burn less of the Earth's stores of fossil fuels. This means we will put less carbon dioxide into the atmosphere so our carbon footprint will be smaller.

That's one way we can help to reduce climate change.

The architect (on the left) is showing the buildings inspector the plans of this new house; the inspector will check that the walls, floor and roof are insulated to the correct standard

Summary Questions

❶ Make a list of the different ways in which a building can be designed to reduce the rate at which it loses heat in the winter.

❷ Explain how insulating a house can help to keep it cool in summer.

❸ A salesman suggests that a house owner could save money by fitting cavity wall insulation. The cost will be £1000. He estimates that heating bills will be £50 less each year. The house owner is not sure – is it worth it?

❹ Many families could save money (and energy) by turning down their heating and wearing warmer clothes. However, most people are reluctant to do this. Suggest why this is.

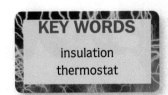
KEY WORDS

insulation
thermostat

How Much Energy?

▶▶ What unit is energy measured in?

▶▶ How can we measure the energy content of our food?

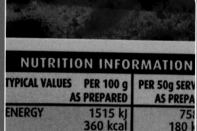

NUTRITION INFORMATION		
TYPICAL VALUES	PER 100 g AS PREPARED	PER 50g SERV AS PREPA
ENERGY	1515 kJ	75
	360 kcal	180
PROTEIN	4.1 g	2.
CARBOHYDRATE	60.0 g	30.
OF WHICH SUGARS	36.7 g	18.
FAT	11.5 g	5.
OF WHICH SATURATES	6.9 g	3.
FIBRE	3.6 g	1.
SODIUM	0.2 g	0.

Foods are labelled to show their energy content, in kilojoules

Did You Know?

An elephant needs to eat for 18 hours of the day. Its food is low in energy, and it's a very big animal.

Be active – use that energy!

Food as fuel

Humans tend to be active. We walk about; we lift and carry things; we run and swim. All of this takes energy.

Our food is our energy source. Foodstuffs contain carbohydrates, fats and proteins. These are all energy stores.

Most packaged foods are labelled with their energy content. Look for this on the label. Two different units are usually shown:

- **calories** (kcal) – the old scientific unit of energy
- **kilojoules** (kJ) – the modern unit

> (a) Look at the cake packet in the photograph. How much energy would you get from eating 100 g?

Just a joule

The standard scientific unit of energy is the **joule (J)**.

- 1 kJ (kilojoule) is 1000 J.

A joule is a small unit of energy. You need about 10 million joules each day, just to keep your body going. That's about 100 J each second – less when you are sleeping, a lot more when you are being very active.

If the energy contained in your food is more than the energy you use through activity, your body will store the excess. It does this by building up fat in your body. This can lead to obesity and heart disease. It's better to become more active than to cut down on your energy intake.

Measuring food energy

Food technologists measure the energy content of foods. To do this, they burn it – just as we burn any other kind of fuel. Then they see how much heat is released.

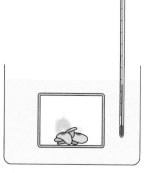

A sample of food is burned inside a closed container; the surrounding water gets hot – the bigger the temperature rise, the more energy it has released

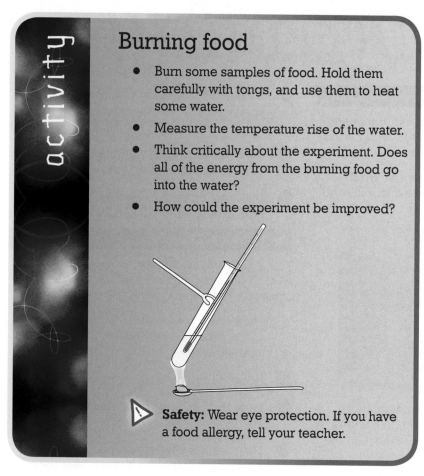

activity

Burning food

- Burn some samples of food. Hold them carefully with tongs, and use them to heat some water.

- Measure the temperature rise of the water.

- Think critically about the experiment. Does all of the energy from the burning food go into the water?

- How could the experiment be improved?

⚠️ **Safety:** Wear eye protection. If you have a food allergy, tell your teacher.

Going green?

A plant gets its energy from sunlight. Its green leaves contain chlorophyll. This captures the energy of the Sun's rays. The energy is stored in sugars and starch within the plant.

If you decided to stop eating and live off sunlight, you would need to grow over 10 m² of leaves – that's as much as a small tree, just to provide 100 J each second. Also you would have problems on a cloudy day and at night!

Summary Questions

1. What units of energy are mentioned on these pages?

2. How many joules are there in 5 kJ?

3. Look at the diagram of the food burning experiment. What result would be found if a bigger sample of food was used? How can this be made a fair test to compare different foods?

4. Some people, known as 'breatharians', claim to live off sunlight and fresh air, rather than food and water. Do you believe this? How would you test their claims?

KEY WORDS

calories
kilojoules
joule

know your stuff

▼ Question 1 (level 4)

Moira's house is on a remote island. Her parents have installed solar cells on the roof and a wind turbine. These generate electricity.

The house also has a diesel generator which can generate electricity when the solar cells and wind turbine are not working.

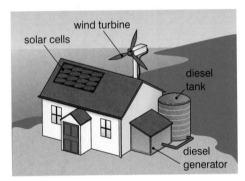

wind turbine
solar cells
diesel tank
diesel generator

a In the table below, for each method of generating electricity (shown on the left), choose the energy resource (shown on the right) which it makes use of. [3]

Diesel generator	Sunlight
Solar cells	Running water
Wind turbine	Fossil fuels
	Moving air

b The solar cells do not work at night. Explain why not. [1]

c The wind turbine does not always supply electricity. Explain why not. [1]

d Explain why it is important for Moira's family to have a petrol generator. [1]

▼ Question 2 (level 5)

Emil is coming down a water slide.

a Name the force which pulls Emil downwards so that he moves down the slide. [1]

b Name another force which acts on Emil as he comes down the slide. [1]

c The water makes it easier for Emil to move quickly down the slide. Explain why. [1]

▼ Question 3 (level 5)

Year 7 are studying energy resources. They made a list of different resources, which their teacher shows on the whiteboard:

geothermal   biomass   nuclear   oil
coal   moving air   tidal
running water   solar   natural gas

a From the list, name three fossil fuels. [3]

b From the list, name three renewable energy resources. [3]

c The class has been reading about a new tidal power station to be built near their town. The local newspaper published this picture of how it will work:

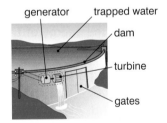

generator   trapped water
dam
turbine
gates

The boxes below show the stages in generating electricity.

A   The generator produces electricity.

B   The gates shut, trapping the water.

C   Water flows out past the turbine, making it turn.

D   As the tide comes in, the water level behind the dam rises.

E   The turbine turns the generator.

Put the boxes in the correct order. The first one is D. [4]

How Science Works

▼ Question 1 (level 5)

Mary and John are investigating a small electric heater. They are using it to heat 200 cm³ of water in a beaker. They measure the temperature of the water every minute.

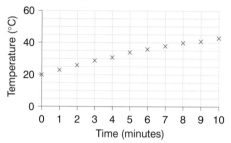

The graph shows their results.

(a) Explain why Mary stirred the water each time before taking its temperature. [1]

(b) John wrote this prediction:

'If we wait twice as long, the temperature of the water will go up twice as much.'

Do the results of the experiment support John's prediction? Explain how you know. [2]

(c) Copy the graph. Add a second line to show the results you would expect if the experiment was repeated with 400 cm³ of water in the beaker. [2]

▼ Question 2 (level 5)

In an experiment to investigate how a spring stretches, Jay and Kay hung weights on the end of the spring and measured its length.

Jay said, 'Every time we increase the weight, the spring will get longer.'

The table shows their results.

Weight (N)	Length of spring (mm)	Increase in length (mm)
0	40	0
1	46	6
2	52	12
3	58	
4	64	
5	70	

(a) Study Jay's prediction and look at the table of results. Was Jay's prediction supported by the results? [1]

(b) Copy the table and complete the final column. [1]

(c) Draw a graph to show the results. [2]

(d) Use your graph, or the table of results, to find out how much the spring stretched for every newton of load. [1]

(e) Kay tried to make a better prediction than Jay. She said:

'If we double the weight on the spring, it will get twice as long.'

What Kay said is not quite right. Write down a better conclusion, based on the results they obtained. [2]

▼ Question 3 (level 6)

Emma carried out an experiment to investigate floating and sinking. She had a wooden ruler. She weighed a lump of Plasticine and then attached it to one end of the ruler. Then she floated the ruler in water and recorded the length of the ruler sticking out of the water.

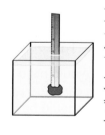 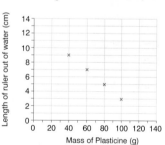

Study the graph of Emma's results.

(a) Use the graph to predict the mass of Plasticine needed to make the ruler sink. [1]

(b) If there were no Plasticine, what length of the ruler would stick out of the water? [1]

(c) Put these materials in order, starting with the most dense: water, wood, Plasticine. [2]

(d) Emma says, 'As the ruler gets heavier, the upthrust of the water gets less, so eventually the ruler sinks.' Explain why Emma is wrong. [2]

How Science Works

Science is never boring - there are so many different ways to find the answers to scientific questions

How quickly does a sunflower grow?

>> How can we find the answers to scientific questions?

Finding the answers to questions (1)

Scientific **enquiries** are all about finding the answers to scientific questions.

On the next four pages you will find some of the approaches you can use … but remember that science can't answer all questions.

There is no one way that scientists work. It all depends on the question they want to answer.

Observing and exploring

'How quickly does a sunflower grow?'

This is an example of a question that we could start to answer by **observing** the sunflower. Just plant a seed, then observe and measure what happens over time.

ⓐ What things could you count as the sunflower grows?

ⓑ What could you measure each week?

Data-logging equipment can help you measure changes over long periods of time.

Researching – using secondary sources

'What type of pollution do cars make?'

You would need to use reference material to find about pollution from cars. You can then use this (and other observations) to explain if we should be worried about the traffic on our roads.

What type of pollution do cars make?

Research a problem

Make up your own scientific question about something that interests you.

Use books, videos, CD-ROMs and/or the Internet to **research** the answer.

Present your question and its answer to display to the rest of your class.

c Why do we need to carry out research to answer some questions?

Classifying and identifying

Sometimes we want to **identify** things, such as plants, animals, rocks, materials or stars. We can use different ways to find out what they are. For example, we can use keys, descriptions or practical tests.

At other times we might want to **classify** something to a particular group. For example, is a material a solid, a liquid or a gas? Is it magnetic? Does it conduct electricity? Or which group does this animal belong to?

Which materials are magnetic?

d Design a key someone could use to identify four people in your class.

Classifying materials

- Classify a selection of materials as metals or non-metals.
- If you could do one practical test to see if you are right, which test would you do?

Summary Questions

1 a) Write a set of instructions explaining to a Year 6 pupil how to research a problem such as: 'Is there any pattern in the number of moons a planet has?'

b) Collect the data you need to answer this question and see what you think.

2 Design a key to help someone decide if a material is a solid, a liquid or a gas.

KEY WORDS

enquiries
observing
research
identify
classify

How Science Works

Finding the answers to questions (2)

▶▶ How can we find the answers to scientific questions?

Here are some other ways to find the answers to scientific questions.

Fair testing – controlling variables

Sometimes we need to carry out a fair test to answer a question like 'How does the length of the string affect the pendulum swing?'

You can answer some questions by carrying out a fair test. Any things that might affect your investigation are called **variables**. In a fair test, you change one of the variables in each test, but keep all the others the same.

Look at the picture on the left.

A fair test will show you how one variable (e.g. the length of the string) affects another variable (e.g. the number of swings in a set time).

On page 156 we will see how to plan and carry out the pendulum investigation.

ⓐ What thing (variable) should the girl in the picture above vary in each test to answer her question?

Pattern seeking – surveys and correlation

Do more weeds grow in the shade or on the sunny parts of the school field?

Sometimes we want to know how one variable affects another, but there may be lots of things that we can't control. We can't keep them the same so we can't really carry out a fair test. These enquiries often involve living things.

The best we can do is to collect lots of data. By increasing the size of the sample we look at, we can have more trust in the data. We say that the data is more **reliable**. Then we can be more certain of any patterns that we find.

If you find some link (or **correlation**), then you should ask yourself why it happens. For example, why do weeds growing in shade have larger leaves?

Looking for patterns

- Carry out an enquiry to see if there is any link (correlation) between the size of a person's hand span and the size of their feet.
- Think about the size of the sample you will test.
- Plot your data on a graph and look for any pattern in the points drawn. Try dividing your graph into quarters. Label the quarters 'Large hand span and large feet', 'Small hand span and large feet', 'Large hand span and small feet' and 'Small hand span and small feet'.

Using models

We can use models to help investigate our questions. For example, we use the particle model to help us understand how materials behave.

The better a model:

- the more things it can help explain
- the more accurate the predictions you can make by using it.

> I find it much easier to answer questions about solids, liquids and gases now I can think about particles.

b Why do we use the model of a fairground log flume when we first learn about electricity?

Using and evaluating a technique or design

In this type of enquiry you are solving a problem using a technique, which might involve a series of steps. For example you might want to find out 'How much salt is in a sample of rock salt?' Or you might be using your science to design a solution to a problem, such as making a burglar alarm.

> Are you sure this circuit could catch a burglar?

c Design a burglar alarm system for a new house. Include a plan of the house and explain how your system will protect the house.

Summary Questions

How would you tackle these questions? Choose from the types of enquiry on the last four pages:

1 How does the temperature affect how quickly sugar dissolves?

2 What is the temperature on the surface of Venus?

3 How can you make your own forcemeter (newtonmeter)?

4 What type of tree is this?

5 Do men have a faster pulse rate than women?

6 Why do gases expand when we heat them up?

7 What happens when we add an acid to chalk?

KEY WORDS

variable
reliable
correlation

How Science Works

The skills of investigation (1)

▸▸ How can we plan a fair test?

▸▸ How can we plan a safe test?

Planning a fair test

Have you ever swung on a rope tied to a tree? If you have, you were acting as a human pendulum. Without thinking about it, you would be finding out the answer to the question: 'What affects how quickly a pendulum swings?'

You can find out the answer more scientifically by carrying out a fair test.

> **a** Make a list of all the things that might affect the way a pendulum swings.

Length of string, mass of the bob at the end, height you release it from, thickness of the string.

Let's see how the length of string affects how many swings it does in, say, 20 seconds

Right, let's sort out all the variables...

Planning a fair test.

We will change:
The independent variable.

We will measure its effect on:
The dependent variable.

We will keep these things the same:
The control variables.

> **b** On a copy of the 'Planning a fair test' sheet, fill in the variables for the pendulum investigation ('How does the length of the string affect the number of swings in 20 seconds?').

First of all, you list all the variables that might affect how the pendulum swings.

This is the start of planning your fair test.

On the planning sheet you will see the term **'independent variable'**. This is the variable that you change in each test. In your investigation it will be the length of the string. All the other variables that might affect the pendulum are kept the same. We call these **'control variables'**.

You will also see the term **'dependent variable'.** We use this variable to judge the effect of changing the independent variable. In your investigation it will be the number of swings in 20 seconds.

Planning a safe test

In any investigation, you must think about safety.

Ask yourself:

- What could go wrong?
- How could it be dangerous?

The hazard might be caused by:

- the way you carry out your investigation
- the equipment you have chosen
- the materials you plan to use or are made in your investigation.

You need to know the hazard symbols below to judge this. All hazardous chemicals will have one of these labels on their container:

Irritant
These substances are not corrosive but can cause reddening or blistering of the skin.

Oxidising
These substances provide oxygen which allows other materials to burn more fiercely.

Harmful
These substances are similar to toxic substances but less dangerous.

Highly flammable
These substances easily catch fire.

Corrosive
These substances attack and destroy living tissues, including eyes and skin.

Filling in a **risk assessment** form will help make sure you cover these questions:

- What is the risk of harm? How likely is it that someone could get hurt?
- Can you change your plan to reduce any risks?
- If an accident did happen, what would you need to do?

Your teacher must check that your plan is safe before you start any practical work.

Toxic
These substances can cause death. They may have their effects when swallowed or breathed in or absorbed through the skin.

> **c** A group decided to test what mass was needed to snap wire made of different metals. What risks would they need to plan for? How might they reduce the risks?

A group were investigating friction.

They wanted to see how the mass in a box affected the force needed to move the box.

1 The title of their investigation was phrased as a question. What is the title of the investigation?

2 What was the independent variable in their investigation?

3 What was the dependent variable?

4 Which variables did they have to control?

Summary Questions

KEY WORDS

independent variable
control variables
dependent variable
risk assessment

How Science Works

The skills of investigation (2)

Planning to collect data

> ▸▸ How can we collect our data?
>
> ▸▸ How can we record our data?

Think about your pendulum investigation again.

You will be changing the length of the string in each test.

> **a** What will you need to measure the length of the string?

You will be counting the number of swings in 20 seconds.

> **b** What measuring instrument will you need to time the 20 seconds?

Which balance would give the more precise measurement of mass?

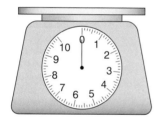

You should think about how **precise** your data needs to be. For example, the length of the string measured with great precision could be 10.100 13 cm. However, for our investigation a ruler with a millimetre scale will be good enough. It can give the same measurement as 10.1 cm (as long as you read it properly!).

> **c** Which measuring instrument can give more precise measurements: a stop-clock which measures to the nearest second or a stop-watch reading to the nearest one hundredth of a second?

Planning to record data

You should also think ahead about how you will record your data.

Scientists record data in tables as they carry out tests.

They usually put the independent variable in the first column. They put the dependent variable in the second column. Any units are put in brackets. So in the pendulum investigation the table would be:

Length of string (cm)	Number of swings in 20 seconds

Some data can be tricky to collect. The data might not be reliable. To help improve the reliability we can repeat readings. We can have more trust in the readings if the repeats are all close together. If one of the repeat readings is very different from the others, you should ignore it and try the test again.

However, if you are doing something wrong in each test, repeating tests won't give **accurate** data. Accurate data are near to the true value you are trying to measure.

We record repeat readings in a table with the second column split up into smaller columns. For example:

Length of string (cm)	Number of swings in 20 seconds			
	First test	Second test	Third test	Mean (average)

You add up the three tests and divide the answer by 3 to work out the mean (average) in the table above.

I don't think that's the type of table we were meant to design.

activity

Carrying out trial runs

Carry out some trial runs of your pendulum investigation. These will help you to decide:

- How long should I make the string? What will be the shortest and longest lengths? This is called the **range**.

- How much should I change the length by between each test?

- How many different lengths shall we test?

- How heavy should the bob at the bottom of the string be?

- Do I need to repeat readings?

- Now you can draw a table to collect your data next lesson.

A group were investigating how the temperature affects the time it takes sugar to dissolve in water. It was difficult to judge exactly when the sugar had completely dissolved in each test. They decided to do their investigation at 20, 30, 40, 50 and 60°C.

❶ Design a table that the group could record their results in.

❷ They asked their teacher for a stop-watch reading to one hundredth of a second to do their timing. Why did the teacher say that the second hand of the clock on the wall was good enough for this investigation?

Summary Questions

KEY WORDS

precise
accurate
range

How Science Works

The skills of investigation (3)

▸▸ How can we analyse the data we collect?

▸▸ How should we evaluuate our investigation?

Now you can collect the data to answer the question below:

'How does the length of the string affect the number of swings in 20 seconds?'

activity

Carrying out a fair test

- Carry out your pendulum investigation, recording the data in your table from last lesson.

Analysing your evidence

Having carried out your investigation, you should try to answer your original question. We need to **analyse** the data collected.

The best way to see any patterns in the data is to draw a graph.

You can draw a line graph because the independent variable, the length of the string, can have any value. You might choose 10 cm, 20 cm, etc. but you could have chosen 10.2, 13.8 or another number. This type of variable is called a **continuous variable**. Your independent variable always goes along the bottom of your graph.

- The dependent variable – the number of swings in 20 seconds – goes up the side.

- You plot the points as small, neat crosses.

- Then you can draw a line of best fit through your mean results: don't join the points like 'dot-to-dot'.

Here are some pendulum data presented on a graph:

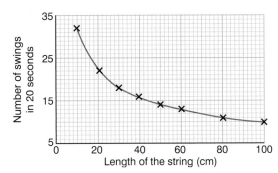

Stretch Yourself

Try plotting all of your repeat readings on a graph. Look at the spread of each set of data. Comment on the reliability of your data.

(a) What pattern can you see from the graph?

In some investigations the independent variable is described by words, not numbers. Then we call the variable a **categoric variable**.

For example, 'Which shopping bag is the strongest?'. The independent variable is the type of shopping bag, e.g. Tesco, Sainsbury, Waitrose etc. There are no values in-between each type of bag. So we cannot draw a line joining points together to form a line graph. We should draw separate bars on a bar chart.

(b) Which type of graph would you use to display the results of the investigation 'Which type of paper absorbs most water?', a line graph or a bar chart?

Evaluating your evidence

As you carry out your enquiries, and certainly at the end, you should always consider the **strength of your evidence**. These questions will help you to **evaluate** your investigations.

- Can you draw a **valid** conclusion using the data collected? Does it answer your question and was it a fair test? For example, you might choose a poor range for the 'Length of the string'. It might be a narrow range of 10.1, 10.2, 10.3, 10.4 and 10.5 cm. This could give data that shows no pattern between the length of string and the rate of swing. Also, was the mass of the bob the same in each test (especially if you did the tests over a couple of lessons)?

- Is your evidence reliable? If you, or somebody else, did the same investigation again would the data be the same? If the data is the same, you have collected stronger evidence for any conclusions you come to. If you have large differences within sets of repeat readings, your data could be unreliable. Then you can't place much trust in your conclusions.

- How could you improve your enquiry? Suggest any improvements you could make. Think of any changes you could make to your method. Explain how they would help.

- If you are doing a 'pattern seeking' enquiry, especially those involving living things, think about the size of the sample you chose. Was it large enough for you to be confident in your conclusions?

Can you draw a valid conclusion using the data collected?

Does the data answer your question and was it a fair test?

Is your evidence reliable?

EVALUATION

Suggest any improvements you could make.

If you have large differences within sets of repeat readings, your data could be unreliable.

Think about the size of the sample you chose. Was it large enough for you to be confident in your conclusions?

Summary Questions

1 Evaluate your pendulum investigation using the questions on this page.

KEY WORDS

analyse
continuous variable
categoric variable
evaluate
valid

Glossary

accurate Describes data that is near to the true value

acid A chemical that can dissolve in water to make a solution with a pH of less than 7

adapted An animal, plant or cell having specialisations for a certain job

afterbirth The placenta, which is pushed out of the womb after a baby is born

air pressure The force per unit area caused by air particles hitting the sides of a container

air resistance Similar to drag; the force of friction when an object moves through air

alkali A chemical that can dissolve in water to make a solution with a pH of more than 7

ammeter An instrument for measuring electric current

amnion The bag of liquid that surrounds and protects an unborn baby

analyse To examine data collected

antacid A medicine to treat heartburn or acid indigestion

anther The male part of a flower which makes pollen

antibody A chemical produced by white blood cells which helps destroy microbes

aseptic Working in sterile conditions

asexual reproduction A method of reproduction which involves only one parent

bacteria (singular bacterium) A type of microbe

balanced Two or more forces whose effects cancel out

ball and socket joint A type of joint, like that in the shoulder or hip, which gives a wide range of movement

base A chemical that will react with (neutralise) an acid

battery Two or more electrical cells connected together

bio-diesel A fuel for vehicles, made from anything which grows

biomass Living material which can be burned to release energy

biotechnology Using biological methods to make useful products

boiling A physical change where a liquid is heated and becomes a gas

breech birth Where a baby is born bottom first instead of head first

brittle Breaks easily

bulb An adapted leaf bud which some plants use to store food

Bunsen burner A piece of equipment used to heat chemicals in the lab

Caesarean section An emergency form of childbirth, where the mother's uterus is cut open to take the baby out

calorie A unit of energy stored in food

carbon dioxide A gas that forms 0.04% of the air

carbonate Part of a chemical that can react with an acid to form carbon dioxide

carpel The female part of a flower made of the stigma, style and ovary

cartilage A tough elastic material which acts as a shock absorber between bones

categoric variable A variable that is described by words, not numbers

cell A single component that provides a voltage in an electric circuit

cell A small structure that makes up most living organisms

cell membrane A thin layer that surrounds a cell

cell sap The liquid found in the vacuole of plant cells

cell wall A rigid layer around a plant cell, which gives it strength

cellulose The material which cell walls are made of

cervix A narrow tube between the vagina and the uterus

chemical change A difference where new chemicals are made

chemical reaction A change, where new chemicals have been made

chlorophyll A green substance found in plants which traps energy from light for use in photosynthesis

chloroplast The part of a plant cell containing chlorophyll, where photosynthesis takes place

chromatogram The results from a chromatography experiment

chromatography A method for separating inks and dyes

circuit diagram A diagram used to show the arrangement of components in a circuit

circulatory system The parts of the body that carry blood

classification System for sorting things into groups

classify Putting things into groups

climate change Long-term changes in the pattern of our weather

colony A group of millions of bacteria, growing on a Petri dish, which can be seen with the naked eye

combustion A chemical reaction that is also known as burning

compass A device which uses a small magnet to show direction

component Any item used as part of an electric circuit

conclusion What has been found out from an investigation

condense A physical change where a gas becomes a liquid

condom A rubber sheath which stops sperm getting from the penis to the uterus

conductor A material which allows electric current to flow freely

continuous variable A variable that can have numerical values

contraception A method of preventing a woman from getting pregnant

contraceptive pill A tablet containing hormones which stops a woman from getting pregnant

contract A muscle contracts when it shortens

contraction When muscles in the wall of the uterus get smaller and squeeze the baby out

control variables These variables are kept constant to make sure an investigation is a fair test

core A piece of magnetic material used to increase the strength of an electromagnet

correlation A link or connection between two variables

corrosion A chemical change between oxygen and metal; rusting is an example of corrosion

corrosive Describes a substance that kills living cells and can eat away at a lot of different materials such as metals

courtship An activity designed to attract a member of the opposite sex

cover slip A thin piece of glass that is placed on top of the specimen on a microscope slide

curd The solid material that is separated from milk in making cheese

cytoplasm The liquid inside a cell where chemical reactions occur

density How heavy something is for its size; it is often measured in grams per centimetre cubed (g/cm^3). Density = mass/volume

dependent variable The variable you use to judge the effect of changing the independent variable

device An item used to perform a task, often run on electricity

diaphragm i) A layer of muscle below the ribs, used in breathing. ii) A rubber disc, used as a contraceptive in women, which stops sperm getting into the uterus

diffusion The spreading out of gas or liquid particles

digestive system The parts of the body used to get nutrients from food

disease Something that stops our body from working properly

dispersal Spreading seeds away from the plant on which they were formed

displacement Movement from one place to another

distillate The liquid collected in the receiving tube in a distillation

distillation A method for separating liquids from each other or solutions

drag The force of friction when an object moves through a liquid or a gas

dye A coloured solution

ejaculation Forcing semen from the penis in sexual intercourse

electric current A flow of electric charge around a circuit

electromagnet A magnet made from a coil of wire, operated by an electric current

electron A tiny charged particle which moves through a metal when a current flows

electron microscope A special kind of microscope, which uses a beam of electrons instead of light and can produce much higher magnifications than a light microscope

embryo The stage in the development of an animal before it has formed most of the main organs

energy What burning fuels release, allowing us to do things

enquiry The way in which we find out the answers to scientific questions

enzyme A protein which controls a chemical reaction in living things

epidermis cells A thin outer layer of cells

epithelial cell A thin layer of cells lining surfaces inside the body, such as lungs, intestines, etc.

erection When a penis becomes larger and stiffer so that sexual intercourse can take place

evaluate To suggest ways of improving an investigation

evaporation A physical change where a liquid becomes a gas below its boiling point

extension Increase in length of a spring or other object

eyepiece A lens in a microscope that you look through

fermentation When microbes use a source of energy and produce a useful product, e.g. beer or wine

fertilisation When the male nucleus joins with a female nucleus

filament A 'stem' which supports an anther

filter A method for separating a solid from a liquid

filtrate The liquid collected after filtration

fixed joint A type of joint, like those in the skull, which does not move

flammable Describes a chemical that burns easily

fetus The stage in the development of an animal after the main organs have been formed

force A push or a pull

forcemeter A device used to measure a force

fossil fuel A fuel formed from materials which were once alive

freeze A physical change where a liquid becomes a solid

friction A force which opposes movement

fruit A structure produced by a plant, which helps the dispersal of seeds

fuel A material burned to release energy in a more useable form

function What something does, its job

fungus (pl. fungi) A type of organism, often microscopic

fuse A component which melts ('blows') when the current flowing through it is too great

gametes Sex cells, such as sperm, egg, ovule and pollen

gas The state of matter in which the particles are spaced out, and move fast in all directions

germination The start of growth of a seed

glands A part of the body which releases a substance, e.g. the liquid added to sperm to make semen

greenhouse gas Any gas (such as carbon dioxide) which contributes to global warming

harmful Harmful chemicals can make you very ill if you eat them, breathe them in or absorb them through your skin

hazard Something that can cause an accident

heating effect The increase in temperature caused by an electric current

hinge joint A type of joint, like the elbow, which moves in one direction

hormone A chemical which carries messages to cells, such as those which control puberty and the menstrual cycle

hydrocarbon A chemical that contains only hydrogen and carbon

hydrogen An explosive gas; it can be tested by putting a lighted splint near it when you will hear a squeaky pop

identify Naming things

image What we see when we look through a microscope

in parallel Components connected side-by-side are in parallel

in series Components connected end-to-end are in series

incubate To keep something, such as microbes, in warm conditions so that it grows or develops

independent variable The variable that you choose to change in a fair test

indicator A special chemical that can be used to find out if a liquid is an acid or an alkali

infectious Can spread from organism to organism

inflexible Cannot be easily bent

inoculation The transfer of microbes onto a growth medium

insulation (thermal) Material used to prevent the escape of heat from a building

insulator (electrical) A material which does not allow electric current to flow freely

irreversible Cannot be changed back

irritant Will make your skin red and itchy

IUD Intra-uterine device used as a method of contraception

joule (J) The scientific unit of energy

kilojoule A unit of energy

lactic acid A substance, produced by bacteria, which causes milk to thicken and form yoghurt

ligament Joins bones together at joints

limewater A chemical used to test for carbon dioxide; if the colourless limewater turns milky then carbon dioxide is present

lines of force Used to represent the strength and direction of a magnetic field

liquid A state of matter where the particles are close together and move randomly

load A force which tends to stretch an object

lubricant A substance which reduces friction

magnetic field The area around a magnet where magnetic materials can be affected

magnetic material Any material which is attracted by a magnet

magnetise To turn something into a magnet

magnify To make something look bigger

mains electricity Electricity supplied from a central source such as a power station

malleable Can be hammered into shapes without breaking

mammary glands The parts of a female mammal's body that produce milk

marrow A jelly-like substance inside bones which produces blood cells

mass A measurement of how many particles there are in a material, it is measured in kilograms (kg)

measure Finding the size of a quantity

melting A physical change where a solid becomes a liquid

menstrual cycle The monthly changes in the reproductive organs of a female human

menstruation The part of the menstrual cycle when the uterus lining breaks down

metal A chemical that is shiny, malleable and a conductor; they often have high melting and boiling points and are sonorous (can make a ringing sound)

method A step-by-step guide to completing an experiment.

methylene blue A stain used on microscope slides

microbe A living organism which is too small to be seen

micro-organism Another name for a microbe

microscope A piece of apparatus used to look at very small objects

mixture More than one substance, not chemically joined together

model A simplified way of explaining observations; models have limits but can be used to make predictions

nerve cell A cell which carries electrical impulses between parts of the body as a method of communication

nervous system The brain and nerve cells, which control the body

neutral A chemical with a pH equal to 7

neutralisation A chemical reaction between a base (or alkali) and an acid; this makes a metal salt and water

newton (N) The scientific unit of force

newtonmeter A device used to measure a force; a forcemeter

non-metal A chemical that is dull, brittle and an insulator; they often have low melting and boiling points

non-renewable Describes energy resources which are used up

nucleus The part of a cell which carries genetic information

nutrient agar jelly A substance which is used to grow microbes

objective The lens in a microscope nearest the thing we are looking at

observing Using your senses to gather data

offspring The young or babies of an animal

organ A group of tissues with a particular job

ovary i) The part of the female reproductive system which releases eggs and produces hormones. ii) The part of a flower which makes ovules

oviduct The tube between the ovary and the uterus where fertilisation takes place

ovulation The release of an egg from the ovary

ovule The female sex cell in a plant

ovum (plural **ova**) The scientific name for an egg

oxidiser A chemical that gives oxygen to another chemical to help it burn better

oxygen A gas that makes up about 20% of the air

palisade cell A leaf cell containing may chloroplasts; the main site of photosynthesis

particle This is what makes up all matter

pathogen A microbe which causes a disease

penis The male organ which puts sperm inside a female

period Another name for menstruation

permanent magnet A magnet which doesn't require a current to make it work

Petri dish A glass or plastic dish with a lid, often used for growing microbes

pH A measure of how acidic or alkaline

physical change A reversible change that does not make a new chemical

pivot joint A type of joint, such as between the skull and vertebrae

placenta A structure in the uterus where substances pass between the mother's and fetus's blood

pole Where the magnetic force of a magnet is most concentrated

pollen The male sex cell in a plant

pollen tube A tube which grows from the stigma to the ovary and carries the pollen nucleus to the ovules

power station A place where electricity is generated on a large scale

precise Describes measurements taken by an instrument with fine scale divisions, for example, 13.23 cm is more precise than 13 cm

prediction A suggestion about what will happen in an experiment, explaining the reason using science

premature birth A birth more than three weeks before it is expected

pressure A force acting over an area; often measured in newtons per metre squared (N/m^2)

properties A description of how a chemical will look and behave

proportional When one quantity increases in step with another

protozoa (sing. protozoan) A type of microbe

puberty The time when we complete the development of our sex organs

pure Contains only one type of substance

random Without a pattern

range The highest and lowest values in a set of data

reactivity A suggestion of how likely a chemical is to undergo a chemical change

record To write down observations and measurements

red blood cell A blood cell which carries oxygen around the body

relax In a muscle, the opposite of contract

reliable Describes data that you can trust

renewable Describes energy resources which will never run out

reproductive system The parts of the body concerned with making babies

research Using sources of information to find out the answers to questions

resistance How much a component opposes the flow of electric current

respiratory system The parts of the body concerned with taking in oxygen and removing carbon dioxide

reversible Can be changed back to what it started as

risk assessment Your judgement of the hazards involved in a task and how to control them

root hair cell A type of plant cell, with a large surface area, adapted to absorb water and minerals

runner A special type of stem which grows across the surface of the soil and on which a new plant starts to grow

scrotum The special bag made of skin which contains the testes

semen A mixture of sperm and liquid

sensitive Able to sense small differences or changes

separating Sort out into pure substances

sex hormones Hormones which control the menstrual cycle and the development of sexual characteristics

sexual reproduction Reproduction involving two parents, male and female

sexually transmitted disease A disease which is passed on in sexual contact

sieve A method for separating different sized pieces of solids

skeleton All of the bones of the body

slide A piece of glass used in preparing microscope specimens

solenoid Another name for the coil of an electromagnet

solid A state of matter where all the particles vibrate and are touching in a regular arrangement

solute A chemical that mixes completely with liquid (solvent) to form a mixture

solution A mixture of a solute in a solvent

solvent A liquid that can dissolve other substances (solutes)

specialised Has a particular job

specimen Something which is looked at through a microscope

sperm cell A male sex cell

sperm tube The tube which carries sperm from the testes to the penis

stain A coloured chemical that makes it easier to see a microscope specimen

stamen The male parts of a flower made of the anther and filament

state symbol A code to explain if a chemical is a solid, liquid, gas or in a solution

stigma The female part of a flower which collects pollen

strong A chemical with a high or low pH number

style The part of a flower which supports the stigma

sustainable Can be used far into the future without damaging the environment

synovial fluid A liquid found in joints and acts as lubrication

tendon Connects a muscle to a bone

testis (plural testes) Part of the male reproductive system which makes sperm

thermostat A device for controlling the temperature reached by a heating system

tissue A group of the same types of cell

trip switch An electromagnetic switch, used instead of a fuse

tuber A swollen underground plant stem which acts as a store of food

umbilical cord Carries blood between the placenta and the fetus

universal indicator A special chemical that can be used to find out the pH number of an acid, alkali or neutral chemical

upthrust The upward force on an object in a liquid such as water, or in a gas

uterus The part of the female reproductive system where the baby develops

vacuole A large space in a plant cell containing cell sap

vagina The part of the female reproductive system where sperms are deposited and where the baby comes out

valid Describes a conclusion drawn from data collected in a fair test, which is designed to answer your original question

variable resistor A resistor whose value can be easily changed

variable Factors that might affect an investigation

vibrate Bouncing around a fixed position

virus The smallest type of microbe

voltage A measure of the push of a cell or battery

voltmeter A meter used to measure voltage

volume The amount of space something occupies

weak A chemical with a pH number between 4 and 10, but not 7

weight The force of the Earth's gravity on an object

whey The liquid left when curds are separated from milk in cheese manufacture

white blood cell A type of blood cell which helps fight diseases

Index

Acknowledgements

Alamy 33.5, 44.1, 56.2, 57.3, 68.1, 68.2, 73.4, 74.1, 78.2, 81.2, 81.3, 81.4, 81.5, 81.6, 81.7, 86.1, 92.1, 94.1, 98.2, 100.1, 100.3, 101.6, 102.2, 108.4, 113.4, 114.1, 116.1, 120.1, 120.3, 121.4, 128.1, 128.3, 130, 131.3, 136.1, 140.1, 140.3, 146.1, 147.2, 148.1, 148.2; **Martyn Chillmaid** 60.4, 61, 70, 78.1, 86, 101.5, 112; **Corbis** 56.1, 66.1, 66.2, 132.1, 134.1, 137.4, 142.1, 144.1, 144.2; **Corel 124 (NT)** 18; **Corel 82 (NT)** 22; **Corel 459 (NT)** 23; **Corel 602 (NT)** 29; **Corel 545 (NT)** 42; **Corel 799 (NT)** 142: **Digital Vision 4 (NT)** 33, 108.1; **DK Images 3.3**; **Mary Evans** 104.1, 105.3; **Fotolia** 100.2; **GreenGate Publishing** 110.1; **iStock** 83.1; **Oxford Scientific OSF** 28.2; **Photodisc 72 (NT)** 88; **Photo researchers** 129.5; **David Sang** 117.4; **SATS papers** 26; **Science Photo Library** 3.2, 5.2, 7.2, 8.2, 9.3, 15.2, 16.2, 16.3, 16.4, 18.2, 19.3, 19.4, 20.1, 24.1, 29.3, 29.5, 32.1, 34.1, 34.3, 34.4, 36.3, 37.4, 38.1, 38.2, 41.5, 42.1, 42.2, 43.4, 44.2, 46.2, 47.4, 48.2, 50.1, 50.2, 54.1, 55.3, 60.1, 61.3, 62.2, 62.3, 62.4, 64.2, 65.3, 68.3, 68.4, 70.1, 91.4, 100.4, 102.1, 106.107.4, 111, 113.5, 114, 122.1, 123.4, 124.1, 125.5, 133.4, 135.4, 138.1; **Andrew Lambert Photography** 58.2, 88.2, 98.1, 102.1

Picture research by GreenGate Publishing and Kate Lewis.

Every effort has been made to trace all the copyright holders, but if any have been overlooked the publisher will be pleased to make the necessary arrangements at the first opportunity.